W9-CKF-449

SWIPE UP
FOR MORE!

SWIPE UP FOR MORE!

Inside the Unfiltered Lives of Influencers

Stephanie McNeal

PORTFOLIO | PENGUIN

Portfolio/Penguin
An imprint of Penguin Random House LLC
penguinrandomhouse.com

Most Portfolio books are available at a discount when purchased in quantity for sales promotions or corporate use. Special editions, which include personalized covers, excerpts, and corporate imprints, can be created when purchased in large quantities. For more information, please call (212) 572-2232 or e-mail specialmarkets@penguinrandomhouse.com. Your local bookstore can also assist with discounted bulk purchases using the Penguin Random House corporate Business-to-Business program. For assistance in locating a participating retailer, e-mail B2B@penguinrandomhouse.com.

Library of Congress Cataloging-in-Publication Data
Names: McNeal, Stephanie, author.
Title: Swipe up for more!: inside the unfiltered lives of influencers / Stephanie McNeal.
Description: [New York, NY]: Portfolio/Penguin, [2023] | Includes bibliographical references.
Identifiers: LCCN 2022061215 (print) | LCCN 2022061216 (ebook) | ISBN 9780593418604 (hardcover) | ISBN 9780593418611 (ebook)
Subjects: LCSH: Internet personalities—United States—Biography. | Social media—United States.
Classification: LCC PN4587 .M36 2023 (print) | LCC PN4587 (ebook) | DDC 302.23/1092 [B]—dc23/eng/20230301
LC record available at https://lccn.loc.gov/2022061215
LC ebook record available at https://lccn.loc.gov/2022061216

Printed in the United States of America
1 3 5 7 9 10 8 6 4 2

Book design by Nicole LaRoche

For my parents,

who always believed I'd write a book,

and made me believe it too.

CONTENTS

1

A wise Instagram caption once said, "One day you will be at the place you always wanted to be." That's how I felt when I first stepped foot in a place I had spent hours staring at on social media: the suburbs of Salt Lake City, Utah.

I know, it doesn't sound like a bucket-list-worthy destination. But to me it was. Sure, I've swiped through photos of the beaches of Santorini and lusted over the breathtaking views from the mountains of Cape Town. But mostly? I have swiped and scrolled and liked and commented on photo after photo taken in Provo, Lehi, Alpine, and Draper. I have heard of their chain restaurants, and even know the specialties to order.

That's because this stretch of highway among the jagged rocks of the American West is the fertile crescent of a slice of the internet that I have spent the past decade of my life obsessed with. In many ways, the world of bloggers and influencers was born here.

It's jarring to visit somewhere you have spent so much time visiting virtually from a phone screen. Scenes look both familiar and foreign. The Utah suburbs are a Candy Land soaked with rugged beauty and dusted with the air of new money and fame. Custom-built McMansions dot the scenery like Chiclets, hanging off the side of rugged hills so majestic and awe-inspiring that they could make even the most devoted skeptic believe in the existence of a god.

In the valleys below, young adults with smooth hair and even smoother skin carry their babies close to their chests as they walk through the shiny strip malls filled with everything you could ever need to live a peaceful suburban existence. This is the land of fast-casual anything; you can get any type of cuisine you wish in a bowl from a counter. You can buy an outfit for a night out at myriad chain stores. You can buy yourself a Diet Coke loaded with heavy cream and a splash of coconut, but it's hard to find a Starbucks or a glass of wine. As foreign as that Diet Coke is though, it's also not, because you have seen video after video of people trying it. After all, it's gone viral on TikTok.

I made my pilgrimage to the land of Mormon influencers in January 2022. By then I had been reporting and writing about bloggers and influencers for more than five years, and following, engaging, and thinking about them for more than ten. Finally, I was in my subjects' natural habitat. Excited would be an understatement.

I had come to Utah to visit one of these influencers after

nearly two years of talking with her about her career and the impact it has had on her life. Shannon Bird had been sharing her life and that of her family with thousands of people on the internet for a decade, and now she was willing to give me a peek behind the curtain, to see what her family actually looked like offline. Instagram versus reality, if you will.

A perky, blond, Mormon mother of five, Shannon is an OG "mommy blogger," the group of women who took over the mid-2010s internet with their version of aspirational parenting. These women, many of whom are Mormon and live in close proximity to Shannon, were some of the first to live their lives for the masses online, make significant amounts of money doing so, and be true internet celebrities, with their lives dissected on gossip forums and blogs.

Shannon isn't the type of mommy blogger you may be envisioning. If the typical mommy blogger posts a beautiful photo of her quiet, serene children coloring in front of a dreamy background, Shannon's kids are on her stories showing off their rabbits (who may or may not be having babies of their own) and roaring around on dirt bikes. If others may show their kids only in perfect outfits with perfect hair, Shannon's kids often look like, well, kids, with messy curls and sometimes just a diaper. And when some moms post idyllic Christmas shots of their brood in matching pajamas, Shannon uploads a video of her husband, Dallin, taking a chain saw to the trunk of their Christmas tree in the middle of the living room, surrounded by her cackling kids.

Shannon is her own person, and thus her own influencer, and she wants to post only what she thinks is actually good content. She could post videos of herself putting on makeup and her kids doing homework in sepia tones, but what fun would that be? If she were to compare herself to anyone, she would say she is the Adam Sandler of mommy bloggers.

"Shannon loves putting out content that the normal lady trying to sell her minivan or better her life is thinking, 'What is this woman doing?'" Dallin, Shannon's husband, explained to me.

I got to know Shannon over a two-year period. Initially, I found her fascinating for one specific reason: much like many people hate and mock Adam Sandler's movies as lowbrow and cheap comedy, many people on the internet mock, belittle, and even hate Shannon. In fact, she has a reputation for being one of the most "messy" bloggers on the internet, with hundreds of pages of comments in online snark forums absolutely ripping her to shreds.

"People think I'm a shit mom," Shannon repeats over and over in our conversations, with the deadpan breeziness of someone who knows she isn't one and doesn't care that people think she is.

The Birds live in Alpine, Utah, a ritzy, upscale suburb that looks like a cross between a ski town and a bougie town populated by Real Housewives and strip malls. As I drove through in my rental car, each home seemed grander and more ostentatious than the last. By the time I pulled up to Shannon's home, which has five bedrooms and is more than fifty-five hundred square feet, it seemed modest in comparison.

Meeting Shannon and Dallin for the first time was surreal, like watching a painting spring to life. Dallin was taller than I expected, greeting me that day with a relaxed surfer bro energy that I would soon learn was his natural resting state. Shannon soon bounded up beside him, carrying her youngest daughter, London, then two, wrapped in a towel. Shannon was glammed up with long bouncy blond curls, wearing a crop top and jeans. I wasn't sure if she had gotten dressed up for me or if this was just how she always looks. After all, she needs to be camera ready a good amount of the time.

Shannon and Dallin live in the epicenter of Mormon bloggerville, surrounded on all sides by some of the most famous names in content creation. Two of the biggest fashion influencers on Instagram can literally look down on the Birds from their massive homes on the hillside above. A major home design influencer lives down the street. And the Birds were willing to show off.

Soon after I arrived at their home, Shannon got an idea. Did I want to go on a tour of her neighborhood, and she could point out all the houses of famous influencers? It could be a true "Utah experience," she said. Um, absolutely. Had I died and gone to Mormon mommy blogger heaven?

Shannon recruited Dallin as the chauffeur for our outing. I rode shotgun and three of her kids, Holland, then eight, Brooklyn, then four, and London came along.

We first drove to the top of a hill near Shannon's house to the home of Rachel Parcell. Rachel started her blog *Pink Peonies* just before Shannon started blogging, and Shannon has known

Rachel and her three sisters, known collectively online as the Skallas (their maiden name), since she was a kid. Now they are all influencers and live in the same neighborhood. Rachel is a major success on Instagram and has turned *Pink Peonies* into an empire, with a line at Nordstrom, her own fashion line, and more than one million followers.

Rachel and her husband, Drew, who works in construction, recently built a massive home that sits on a hill literally looking over Shannon's. I'm not sure if I should be embarrassed to admit this, but when we drove by I recognized it right away from her Instagram-perfect Valentine's Day wreaths framing the front door.

Dallin nodded to it. "There's her house," he said, "and there's her dog," a yellow Lab relaxing on the grass in the sun. I felt weird taking a picture, but the Birds egged me on, so I did. Of course, I texted it to my friends, asking if they could guess whose house this is. They knew immediately. "This is unreal!" one wrote back. From Rachel's house, we drove down the street to see where her sister, Emily Jackson, also an influencer, is building a similarly enormous house.

From there, we zipped down the mountain, as Dallin pointed out to me all the other enormous mansions and whom they belonged to. Alpine used to be a relatively modest town, he explained, but Salt Lake City has experienced a tech boom in recent years. Overstock.com and Ancestry.com are just two major players in tech that are headquartered in what some now call the "Silicon Slopes." Since they moved in, the Birds have watched small home after small home get knocked down to be replaced by enormous McMansions.

When I mentioned to Dallin I was surprised by how many tech firms were in Utah, he pointed out to me that startups weren't the only innovative new business driving money to where he lives. Influencers are also tech startups, driving a remarkable amount of wealth into Alpine and the surrounding communities.

We took a tour break for a snack at Swig, a shop selling the aforementioned viral soda. Swig recently had gotten big on TikTok because of its unique business model: it sells soda concoctions to Mormons in the style of coffeehouse drinks or cocktails. Most Mormons don't drink hot drinks or alcohol, so places like Swig have popped up to fill the fancy-beverage void. They have become massively popular both in real life and on social media, where non-Utahans comment about their perceived oddity, then try them for themselves or post videos about their favorite drinks they have tried when they visited.

I got a "classic" drink on the menu called "The Founder," which consists of Diet Coke, lime, sugar-free coconut syrup, and coconut creamer. It tasted exactly like a Malibu rum and Diet Coke I would have drunk in college, which I announced to the delight of the Birds.

Sodas, sickly sweet sugar cookies, and pretzel bites in hand, we headed out to finish our blogger tour. I was probably most excited to see the home of the mega home influencer Shea

McGee, of Studio McGee (I own several items from her line at Target), which was beautiful but more understated than I expected.

As I stared at her house, I thought about my house, which had been so heavily influenced by hers. Did I have a personal home decor style before I started following Shea on Instagram? I don't really remember, but I know that as I looked through photo after photo of Shea's style, mine came into focus. I could say the same about so many things in my life.

When I wanted to figure out an outfit for a night at the bars in my twenties, I would scroll through bloggers' outfits on Pinterest. When I wanted to learn how to cook something, I'd consult a food blogger. When I realized I had too much credit card debt, I would read financial diaries. When I wanted to improve my marathon times, I read race reports on running blogs. When I got married, I went to every single blogger I had ever followed who was married and read their entire wedding series. When my husband and I decided to buy a condo and I was filled with so much anxiety I thought my head might explode, I found comfort in watching Instagram Story highlights from influencers who described their own home-buying experiences. When I dealt with a year and a half of infertility, I found similar comfort in reading in-depth posts from influencers who had struggles just like mine. When I was pregnant with my daughter, I devoured Instagram posts and Stories like "first-trimester recap" and "what I put on my baby registry" with the fervor of a college student cramming for an exam. I read a few baby books, but most of my preparation came from these women online, guiding me through.

As I sat in Shannon's car, I began to wonder, *Who would I be without them*?

My fascination with influencers probably arose because of one rather embarrassing fact about myself. I am nosy.

Come on, don't tell me you aren't too. It's one of the fundamental reasons the influencer industry exists. Isn't there something amazing about getting an inside peek into someone else's life? It's enthralling to watch a person who is willing to cut their life open like ripe fruit and then, hands outstretched, invite us all to consume it. Especially if we view them as more conventionally attractive or powerful than we are.

This fascination is why I write about internet culture as a journalist, and specifically about influencers. It's not a topic I ever expected to write about. In college I thought I'd become a "serious" journalist, reporting on Capitol Hill or, I don't know, writing about the UN or something. But once I became a journalist, reporting on what I had always viewed as more serious pursuits didn't interest me. Because when I wasn't at work, I spent my time consuming influencer content.

It was the voyeur in me. I am endlessly fascinated by influencers: how they live their lives so publicly, the things they are willing to share, and how they navigate the internet on their own terms. Some made me want to buy things from them, others made me mad, and still more made me cringe, but they all held my attention. When I talked to my good friends, they were all following their own influencer obsessions. But I began to realize that no one was really talking about them in the mainstream media.

In 2016, as a reporter for *BuzzFeed News*, I began to write about influencers as part of my day job. When I began to seriously cover them, I noticed a few things I found interesting.

At the time, people in popular culture tended to throw around words like *influencer* and *content creator*, and did so interchangeably. But these terms refer to two distinct groups of people. The first group is what I in this book call "video creators," or "creators," as others in the media have called them.

This industry was born on YouTube. Since the first videos started going viral in the late 2000s, the medium has become a hotbed for virality and social media fame. The earliest video creators pioneered an entirely new industry by creating easily consumable short-form video entertainment. The medium soon became known for a variety of tropes, like daily video diaries known as vlogging, comedy skits, and stunt or prank videos, which thousands have tried to emulate in a quest for their own social media fame. In 2012, an app called Vine made stars of even more of these video creators, who were able to successfully turn the app's six-second looping video clip format into full-on comedy shows. Once Vine shut down in 2016, many of these "Viners," like the Paul brothers, Lele Pons, and David Dobrik, were able to successfully migrate their followings to YouTube.

Now, of course, the next generation of these types of video stars has been born on TikTok, the newest app on the scene that relies on video content as the main form of entertainment. Whether on YouTube or Vine, these video creators are akin to actors working in television or film.

The second distinct group of people making money from content creation on the internet are what I call "influencers" in this book. These creators primarily communicate via text and photos—think blogs, then Instagram photos and captions—and their content is akin to a lifestyle or fashion magazine. While they do post video content, like Instagram Stories, Reels, or TikToks, their content is different in a few distinct ways.

Video creators often sought to shock and entertain, through increasingly more daring stunts or pranks, or viral comedy videos. The goal was to continue to make viral videos that would run their numbers up the charts of YouTube and get them more and more subscribers. Even creators who focused on beauty, fashion, or lifestyle on YouTube primarily got big due to their videos' virality, and many continued to make big splashes through "drama" with other YouTubers.

Influencers, in contrast, sought a more sustained relationship with their audience. Similarly to a magazine or lifestyle website, bloggers sought to cultivate an audience who would keep coming back and would check their websites weekly if not daily for updates and inspiration. The focus was less on virality and more on consistency. Now they do so through a variety of mediums, including blogs and Instagram posts and Stories, but also newer forms of expression like TikTok and popular newsletter platforms like Substack.

If you want to think about it another way, influencers strive to create a sustained and stable flow of content that inspires loyalty from their followers. Eventually, this loyalty converts into influencers being able to sell their audience products based on their recommendations. If we go back to comparing creators to established forms of media, influencers are akin to magazine editors.

For years, I had been obsessed with bloggers, who then slowly began to pivot somewhat, if not entirely, to Instagram. Other reporters, though, seemed to focus on video creators, and I began to realize they were getting the most attention in the mainstream press. At the time, major YouTubers like Grace Helbig, Lilly Singh, and Hannah Hart were beginning to get some serious attention in Hollywood. Even second-tier YouTubers had millions, if not tens of millions of followers on Instagram, which was not even their primary platform. By contrast, most influencers topped out at around a million Instagram followers, with many who had been in the game for a while sitting somewhere between three hundred thousand and five hundred thousand. These influencers were successful in terms of getting brand deals and making money, but it seemed like no one was paying attention. (In fact, until quite recently, much of the money in the influencer marketing industry was specifically going to campaigns on Instagram, where influencers reign.)

It was a little confusing at first. I was obsessed with bloggers and Instagrammers, and so were many of my friends, who didn't work in the media industry. But in mainstream media, it was almost like they didn't exist. Why was that?

An easy explanation is to look at the platforms. Instagram didn't prioritize getting mainstream Hollywood deals for its creators as YouTube and TikTok did. Video platforms are also just plainly more visible, and thus more profitable. A YouTuber who gets a Netflix show is going to be a bigger deal than an Instagram influencer who starts a fashion line at Nordstrom.

I do think, though, that other factors come into play here. The most glaringly obvious is who runs the influencer industry, and to whom it appeals. According to the market and consumer data research firm Statista, as of 2019, 84 percent of the influencers creating sponsored content on Instagram were women. Anecdotally this seems obvious, and most of the influencers on Instagram are also creating content revolving around traditionally female pursuits, like fashion, parenting, beauty, or just general lifestyle.

While many YouTubers are also women, and some of the biggest, like beauty and family vloggers, fall into these same categories, there are many more famous men who have emerged from YouTube than have from blogs or Instagram, and the most popular content on YouTube, such as pranks and comedy, has broader appeal.

Unfortunately, the type of content influencers create tends to rank low on the respectability index. Many people in American society either don't care about what influencers create or actively loathe the types of media they put into the world. Fashion is considered wasteful and conspicuous consumption, beauty is considered vain, and mommy blogging, even by culture writers who consider themselves bastions of feminist ideals, is considered pitifully backward at best and actively harmful at worst.

In early 2022, I surveyed approximately twenty influencers on the current state of their careers and how they felt about them (they did so anonymously, to be as candid as possible). Some had half a million or more followers and some had fewer than fifty thousand. They represent a multitude of ethnicities and come from a variety of backgrounds. One of their biggest complaints was the struggle they face to be taken seriously, and what they feel are misconceptions about the industry from the public.

As one influencer told me: "I think it's rooted in misogyny from such a female-dominated industry. They can't fathom that thousands of women have made full-time careers sharing things they love with flexible work hours. Just because I'm not breaking a sweat every day doesn't mean this job doesn't take skill. They would never approach someone who works a desk job at an ad agency the same way, and we're most likely using the same skill sets."

Caitlin Covington has been blogging about her personal style and life since 2012 and is one of the pioneers of the industry. For many years, she was proud to call herself a blogger. But over time, she watched as the perceptions of her work grew more and more tainted, because of both mischaracterizations in the media and bad apples that hurt the industry's reputation.

"I think people view influencers as really narcissistic and

materialistic, and I just dress up and I'm so egotistical I only post pictures of my face on Instagram," she said. "I just hate feeling like that, because that's not who I am, and I just worry that people will think that about me."

When she tells people what she does, she sometimes feels embarrassed.

"I really hate to tell people that I'm an influencer now," she said. "I'm like, 'Please don't judge me, I'm an influencer.' I feel like it's less respected now. People know the influencers can move products and really influence people. But it's just not a respectable career because there's so many negative connotations that go along with it."

I don't want to imply that the industry is perfect or deserves no criticism. Influencing can be harmful to the mental health of both the influencers themselves and the followers who consume their content. In 2021, *The Wall Street Journal* reported on an internal study by Meta (which owns Instagram) that found that 32 percent of teenage girls said that "when they felt bad about their bodies, Instagram made them feel worse."

I think very few women were surprised. There's no doubt in my mind that scrolling through images of women who look prettier, more fashionable, more fit, and more put together than me on a daily basis is bad for my brain and self-esteem, and it's a difficult problem to fix.

Additionally, there are many other problems with influencing as a career. It can be exploitative of children, who have no say as to what their parents share about them with thousands of people, and have no legal protections ensuring their rights aren't being violated. It is a systemically racist industry, with a huge pay gap between Black and white creators. The industry also often rewards those who cheat to get ahead through schemes like buying followers or inflating engagement. And it can warp the influencers themselves, turning them into essentially product-shilling robots, who are willing to do or say or link out to anything to make a quick buck.

In order to fully analyze and criticize something, though, you must take it seriously as something worthy of criticism. For too long, the influencers many women care about and follow the most have been mostly absent from the public discussion about the creator economy or simply written off as vapid product pushers. At first, the only people who really seemed to pay attention or hold the influencers accountable were snarkers on online forums. Nowadays, the stereotype of an influencer in mainstream media is an egotistical, brain-dead woman posting lackluster outfits on Instagram due to pure narcissism. There's been a lot written about the business and culture of YouTubers and now TikTokers, but comparatively little real analysis and critique of influencers as their own separate entity. It'd be like if our entire entertainment media ecosystem wrote only about television but ignored movies, or wrote about rock music but ignored pop. Influencers are different, and so is their impact on culture.

It's time for more nuance. That's why I wrote this book. In these pages, my goal is to bring the multibillion-dollar influencer industry to life as its own entity and to explore both what it is like to have this unique and complicated career and how the industry as a whole has impacted an entire generation of young women. The influencer industry is also majority female-owned and -consumed, and has created an astonishing number of female millionaires in a very short amount of time.

And if you want to understand mainstream female culture today, you should examine how influencers have contributed to it. I am just one example of someone influencers have had a chokehold on over the past decade. They have shaped how I decorate my house, how I shop for clothes, how I think about decisions relating to parenting and finances. Maybe they have even shaped my worldview on serious issues like politics and relationships, without me realizing it. Now, multiply me by millions of women all over the country. What does it mean that this cohort of women is driving so much of our purchases, our thoughts, and our lives? And how does it impact them?

To tell this story, I followed three influencers from different corners of the industry, shadowing them for two years as they lived their lives online. The first, the aforementioned Caitlin, is who I would call the most stereotypical influencer. Caitlin Covington is a fashion and lifestyle blogger who rose to fame based on her aspirational aesthetic and girl-next-door wholesomeness. When you think of an influencer, you probably think of someone like Caitlin.

But there are other types of influencers too. Mirna Valerio, a teacher, attracted fans and media attention as a blogger who wrote compelling narratives about her experiences as a "fat girl running" when she wasn't at her day job. She has been able to turn that into a full-time influencer career that has made her more money than she could have ever dreamed of. Mirna doesn't do a lot of the stereotypical influencer things (you won't find her posting affiliate links to her dresses), but she represents an important part of the influencer industry that is rarely discussed. Influencers who appeal to a certain niche or genre, like running or appreciation of the outdoors like Mirna, are nothing like the stereotype of an influencer but are some of the industry's most successful players. As a Black plus-size woman, Mirna not only has been able to make money but has been able to enact change at the highest levels of the athletic apparel industry and make more of an impact than she could have dreamed of.

And then there's Shannon, an influencer who represents several important parts of the industry. She was one of the original "Mormon mommy bloggers," a cohort of women who defined and shaped the industry as we know it today, and whose success created fraught questions about mothering in public. Shannon also has for many years been facing a barrage of criticism on online snark forums. Indeed, she may be called one of the main influencers they love to hate. Shannon both reviles and has tried to embrace this label, with interesting results for both her personal and professional lives.

My goal in the following pages is for you, the reader, to actually get to know these women and everything you ever wanted to know about the secretive influencer world they made. You will learn how the industry actually works (yes, some influencers hire people just to respond to their DMs) and everything you have ever wanted to know about what it really feels like to be an influencer. It's often said that Instagram is just a highlight reel of one's life, but in this book, I'm going to show you the outtakes.

Swipe up to read more. I'm kidding! Just turn the page.

2

I am standing in a pristine forest in Vermont on a beautiful September day. Though the air still has the slight mugginess of late summer, there are already yellow leaves on the ground beneath my feet. The gargantuan trees surrounding me reach high into the sky, blocking out most of the light from the sun.

Until I climb a little higher. As I crest a small hill among the trees in the woods, I see a clearing before me, bathed in the golden afternoon light. Small particles float in the glow like fairies under a spotlight. For a minute I almost feel like I have stumbled upon one of those secret places in the books I devoured as a child, in which an ordinary girl accidentally travels through a portal into some magical land.

Beside me, a voice breaks my reverie. "This is really fucking cool."

I turn to face Mirna Valerio, who stands with a huge grin. I am standing on this parcel of land because of her. This land is Mirna's dream.

Mirna has been working for years for this moment. Here, in rural Vermont, she has found her place. After growing up in Bushwick, Brooklyn, and then spending most of her adult life hopping from state to state to teach at a series of boarding schools, Mirna has decided to put down roots. Now she is planning to build her dream home on a parcel of untouched land just like this one and invited me to tag along.

Mirna is a dynamic person. When you're around her, you want to listen to whatever she has to say. Her personality and energy seem too large for her shorter frame; when I first met her, I was surprised that I was much taller. Her eyes dazzle when she speaks, and she doesn't break eye contact. I think she could convince me of anything in five minutes. She is full of big, ambitious ideas, giving people confidence and inspiring them to live as joyfully as she does. And she makes you believe she can follow through. Her website declares she has "focused her life on spreading health awareness, promoting diversity and inspiring and uplifting others to LIVE LARGE and be in charge of their own happiness!"

Mirna is a natural influencer. But if I asked you to picture an influencer, you probably wouldn't imagine Mirna. The popular influencer stereotype is young, pouty, leggy, conventionally sexy girls, who post photos of their clothes and bodies on Instagram and get way too much money, fame, and free stuff in return. They sell random stuff through affiliate links, which pay them every time someone buys something they recommend, and they will shill anything for a dollar. They are shallow and have little purpose besides attaining what they deem to be physical perfection and money.

Mirna is not that. She's older than most influencers, in her mid-forties with a teenage son. She's plus-size (she just says "fat") and Black. She's not a fashion blogger; she's a running blogger.

Well, she used to be a running blogger. Now she is much more. Mirna does all kinds of things she never used to do, like skiing and mountain biking. She's a full-time content creator and is sponsored by some of the biggest names in professional sports. She gets to do and see things that she never could have imagined growing up. For most of her childhood, she had never spent more than a night or two outside New York City. Now she spends hours riding through trails, racing down snow-covered hills, and just being outside.

Mirna doesn't have millions of followers, but her more than one hundred thousand followers are extremely loyal. Many have been following her for a long time, and when Mirna recommends something to them, they pay attention. They feel like her success is their success too. When Mirna became a global ambassador for Lululemon, even gracing a billboard for the brand in her hometown of Brooklyn, her fans' excitement was palpable. To them, this wasn't just their favorite blogger getting a cool modeling gig, it was a movement.

"Your work is huge, changing the fitness narrative, how we

see ourselves and how the world sees people," commented one follower.

"If I ever decide to get anything from lululemon, it will ONLY be because of YOU," wrote another.

Thus, Mirna has become extremely valuable to the brands she works with, and in the process, she has been able to create a whole new career for herself.

Before she became a full-time content creator, Mirna was just your average teacher. Well, that's not true. Mirna has never been able to be defined by just one thing. She is a runner, biker, and skier, but also a trained musician (she graduated from the Oberlin Conservatory of Music in 1997 with a dual degree in Spanish and vocal performance after attending classes at Juilliard as a high schooler; she also plays the piano). She has written a memoir and is currently writing her first novel, about a high-powered Black woman who reconnects with herself and deals with childhood trauma through time in nature.

This frenetic energy makes Mirna unique, but it has sometimes been her downfall. She spent her twenties and early thirties hustling at a truly astonishing pace, trying to make sure she provided the best possible life for herself and her family, but sacrificing her own health and wellness to do it.

Mirna first fell in love with running as a teenager when she was in boarding school in Westchester County and picked it up to get better at the sports she was playing. She continued her active lifestyle through college and beyond when she worked in corporate America in New York City. After she got married and had her son, she switched careers to teaching, first at a school in Maryland.

At the time, she commuted up on the weekends to see her husband in New York, which added stress and took away free time. Being active and healthy fell by the wayside, and Mirna remembers both she and her son were constantly sick. She decided to take a job at a boarding school in New Jersey to lessen the stress of the commute, but that job was even more stressful, as she was living in and running one of the dorms. On top of that, she was still commuting to Maryland to teach piano lessons and voice lessons, and for Spanish tutoring. Oh, and she was also in grad school.

"I just could not get well," she said, remembering having a sinus infection for, like, "two years."

All the stress caught up to her. One day in 2008, when she was thirty-three, she was driving when she began experiencing chest pains. Mirna was terrified but feared the hospital bills from an ambulance ride more than she feared what could be happening. She drove home, and a friend drove her to the hospital instead. After hours of poking and prodding, a doctor told her that while they were worried about a large amount of inflammation in her body, Mirna wasn't having a heart attack. She was having a panic attack.

Mirna blanched. Black people don't have panic attacks, she told the doctor. But of course, she says now, she was panicking.

She was pushed to her limit. At a follow-up appointment with a cardiologist, the doctor told her what she was doing wasn't sustainable. "How old is your son?" she recalls the doctor asking (he was five at the time). "Do you want to see him grow up?" It was the wake-up call she needed to start to prioritize herself, and she started to run again for both her mental and physical health.

"Running . . . gave me my life back," she said.

I first heard of Mirna ten years after that wake-up call, when she was featured in *Runner's World* magazine in 2018. The subhead of the article asks: "Is it possible to be fat and fit? Healthy and happy as well as heavy? At 250 pounds, distance runner Mirna Valerio provides an inspiring example."

In the accompanying article, Mirna describes the fascination the larger running community has with her, as an outlier in the sport. "People always say to me, 'Anyone who runs as much as you do deserves to be skinny.' Of course, what they're really saying: 'If you do all this running, why are you still so fat?'" she says in the piece.

I've been an avid runner since I was fourteen, and my hobby created some of my earliest obsessions in the influencer space. For a time, I devoured content from running bloggers, who, like Mirna, posted painstakingly about their training plans, their race recaps, their injuries, and their gear. The women were almost uniformly thin and mostly white, with muscled yet willowy legs and hip bones jutting out of tight running shorts. Their race times were ungodly fast, and while I enjoyed reading them, they also gave me a small inferiority complex. Why should I feel good about running my best time in a 10K when that blogger can do it in half that?

Mirna was the antidote to these other running bloggers, which is how she got noticed by publications like *Runner's World*. Mirna doesn't run fast and she's not trying to break any records. She started a blog, *Fat Girl Running*, in 2011 after getting positive feedback on her long, involved Facebook posts about her experiences running road races. She didn't post for attention. Running made her feel alive, and she wanted to celebrate that. In the mid-2010s, when the only running apparel ads I ever saw were yelling at me to push to the limit or, you know, Just Do It, Mirna offered a new reason to run, or to do any sport: for health and mental wellness.

This concept was pretty revolutionary at the time, and companies and the media were starting to catch on. Mirna was and is more than happy to be their poster child. Her breakout moment came in 2015 when she was featured (alongside a then-not-as-well-known Ashley Graham) in a *Wall Street Journal* article titled "Weight Loss or Not, Exercise Yields Benefits."

Mirna has no idea how the *Wall Street Journal* reporter found her blog. It's not like she was super well-known or had a huge readership. (As a journalist, I can speculate that the reporter had an assignment, googled people who fit the description, and Mirna's blog happened to come up. Just a guess.)

The reporter contacted her completely out of the blue. Mirna was surprised but more than happy to help, telling the reporter all about her background and philosophies on exercise and wellness. The resulting article even plugged her blog, saying it "includes recommendations on plus-size workout gear and a post on her frustration at doctors assuming at a glance that she is inactive."

"And then, bam, that was the start of my social media career," Mirna said.

At that time, she was working as a teacher at a boarding school in rural Georgia. In typical Mirna fashion, she wasn't doing just one thing. At that school, she served as chorus director, Spanish teacher, and cross-country coach, as well as the school's director of equity and inclusion.

For a few years, Mirna kept doing her day job while slowly building her new career online. She has been in high demand. She has never reached out to a brand on her own because they have never stopped calling.

In 2017, she made five thousand dollars in one day for a shoot for JCPenney, a payday that made Mirna realize that this wasn't just a side hustle. She could make real money here, more than she ever could teaching, and more important, she could make an impact. That same year, a doctor had messaged her on Twitter. He had seen a video she was in that REI had produced and asked if she would be willing to come to speak at a conference he was planning and "just tell her story." From there it snowballed through word of mouth and Mirna was asked to do more and more speaking engagements. Over time she developed different programs and speeches on self-love, confidence, and body image.

In 2017, she published her first book, a memoir titled *A Beautiful Work in Progress*, all about her story and her journey to body confidence and a love of running.

At her day job at the boarding school, colleagues joked with Mirna when she took off on this or that opportunity that she was becoming a star. But the final push she needed to actually take the leap of faith to pursue her new career full-time came from an unlikely source. One day, a student whom she didn't even know approached her and told her he had seen what she had been doing. He told her he thought she should leave the school and pursue her influencer career full-time, because, she recalled him saying, "Your star is rising." Mirna was stunned, but it sealed the deal. Soon after, she made an appointment with the head of the school to resign. He told her he had known this day would come because so many media outlets had been coming to the school to film her.

Mirna took a chance on herself and quit her job at the school. After an eighteen-year teaching career, she still sees herself as an educator, but now she's educating the world through her influencing, her writing, and her speaking. A perk of the new role is she can do it from anywhere. After spending some time living with her parents in Brooklyn to weigh her options, she decided to check out the Green Mountain state. She moved in for a few months to a bed-and-breakfast in downtown Montpelier to try it out.

Immediately, she was hooked. To an outdoor enthusiast like Mirna, Vermont is paradise. Her teenage son, who had been away at boarding school, as Mirna had done, soon asked if he could finish high school with her, and the two of them got an apartment overlooking the main street in the quiet capital city. Her son, Mirna said, is doing amazingly well in his new home.

In Vermont, the burnout that plagued Mirna as a young woman seems to be gone. In fact, she seemed to have endless energy, greeting me with an exuberance you rarely find at seven thirty a.m., when I knocked on her door to begin our day.

Mirna led me up to her apartment, which peers out onto Montpelier's small and charming downtown. The spoils and hallmarks of her life as an influencer were everywhere I turned. A desk in the corner prominently displayed a ring light for content creation, and the table next to it was piled high with different brand-new pieces of clothing from one of her sponsors, Lululemon (I know, right?).

As we walked out the door to film some content, Mirna, wearing Lululemon, grabbed her Hydro Flask bottle (also a sponsor). I joked that I was also on theme because I was wearing Lululemon leggings as well and had a Hydro Flask with me, and just happened to be wearing Merrell (another sponsor) hiking shoes. Of course, I had to pay for it all.

We walked downstairs and climbed into Mirna's new Ford Bronco. Ford is, you guessed it, also a sponsor. "Filled with gratitude that I get to enjoy wild, spectacular beauty all over, but this state here is really special. Stay tuned for some classic autumn in Vermont content next month! #teambronco_ambassador," she wrote in a recent #sponsored post for the car company.

Mirna doesn't usually do campaigns in exchange for free stuff without also getting paid, but for the free car, she made an exception (she also gets gas cards from Ford as part of the deal). It allowed her son, then seventeen, to drive her other car because she now had two, which is great. Before she made the deal with Ford, she had been talking to Volvo about a similar deal, but Ford's terms, getting a car for free in exchange for creating content for social media, eventually won out.

After they made the deal, she got a call one day from a random man. He was from Ford, he said, and he was driving her sponsored Bronco to Vermont as they spoke, all the way from Detroit. He would be there in a few hours. Mirna was a little surprised but just said, "Okay, cool!" When the man arrived, she asked him, "So . . . how are you planning on getting to the airport?" "Oh," he replied, "I was just going to call an Uber or a taxi service."

But there was no real Uber or car service or taxis where she lived. So she drove him to the airport. They had a nice chat.

The fact that a girl from Bushwick now lives in a place where you can't even call an Uber isn't lost on her.

As she once told me, "This is clearly where I need to be."

GROWING UP, CAITLIN COVINGTON felt like she had the world at her fingertips.

She was raised in Charlotte, North Carolina, in a family out of a Norman Rockwell painting. Like many millennial children, she felt she had no limits to what she could achieve. Young white girls growing up in the 1990s were pumped up with girl-power mantras, confident that any of the restrictions they may have faced in the past were a relic of the old sexism of the past millennium (of course, the reality was just a little more nuanced than that). Caitlin believed it. All she had to do was go out and grasp her destiny.

What Caitlin wanted was to be a tastemaker. Fashion intrigued her from an early age. First, she fell in love with Limited Too. The '90s and 2000s fashion staple was all the rage and Caitlin was obsessed. Her parents were comfortable but there wasn't unlimited money for an entire Limited Too wardrobe. So, Caitlin spent all of her allowance on clothes, saving it all for one sweet rush of pink-and-sparkle purchases at the mall, even if that meant she had no money left over for anything else.

It wasn't just about looking cool or trying to flex. Fashion was a way for her to express herself to her peers at school and on the weekends, to show others the creative workings of her inner mind. She loved the feeling putting together an outfit gave her, the rush of satisfaction she felt when she saw everything come together in the way she had imagined it. She loved the confidence she felt slipping it on, heading out into the world, and putting her best foot forward. When people saw Caitlin, they saw what she wanted them to see.

As her interest in fashion grew, so did her love of magazines. She loved to write, as well; as a child she kept a prayer journal, writing long missives to God. As a tween, Caitlin combined her interests and made her own magazines on her home computer.

She designed "style guides," and then printed them out just as if they were *Vogue*. Flipping through the glossy pages, Caitlin saw her future. This is what she could achieve if she just worked hard enough.

After high school, Caitlin enrolled at the University of North Carolina at Chapel Hill. Everything she did was focused on her eventual goal. She interned with a Charlotte news station and a local women's lifestyle magazine and even did an externship with a real New York lifestyle publication. She didn't get to actually go to New York because it was all online, but she was getting closer.

However, all the work experience and classes didn't feel like enough as her graduation neared. Caitlin could imagine the thousands of young women just like her who would be vying for the same entry-level magazine job. By 2012, the business of magazines was struggling even if the glamour remained. The news industry had been growing more and more volatile for years, with ad revenue shrinking as Facebook and Google gobbled up larger and larger shares. Local newspapers were folding, and magazine powerhouses like Conde Nast were tightening their belts. What else could she do, she wondered, to show future magazine bosses how passionate she was?

One day, it came to her: a blog. Caitlin didn't follow too many blogs herself and didn't know of any that were focused only on fashion, like the digital version of the style guides she had created on her childhood home computer. What if she started a blog and made it like her own little online magazine? She could show off her skills, her style, and her writing, and then show it to potential employers like a portfolio of what she could do.

Caitlin had no idea how to create a blog, so she winged it. When she realized she had to name her website, she decided that she wanted people to be able to go to her blog and immediately understand what they were going to get. So, she came up with a way to describe herself. One: she is Southern. Two: she had long curly hair. Three: at the time, she loved pearls. Thus, *Southern Curls & Pearls* was born. (If she were to create a blog today, she says, she would never name it this. But now it is an indelible part of her.)

"Hello blog-world!" she wrote in her first-ever post. At the time, she was twenty years old. "My first post . . . how exciting! I've been following blogs for a while now, and just decided to try my hand at it. Wish me luck, I'm still getting the hang of things."

Here were the things Caitlin wanted readers to know about her. She was a "daughter, a sister, a journalist, a writer, a fitness and health enthusiast, a southern gal." Her mom was her best friend. She loved frozen yogurt, the beach, her sorority (Kappa Delta), and shopping. She was "pretty artsy." Her planner "contains my life." Subsequent posts detail her senior sorority rush and the family party for her twenty-first birthday at her favorite restaurant, P.F. Chang's ("sesame chicken, anyone?" she wrote).

"I can't really explain why people were so interested or why it grew so fast," she says now of her blog, "other than I was just kind of a normal, everyday girl. But I made it look prettier."

I THINK I KNOW. If you were to ask most people to picture the epitome of an influencer, many would say Caitlin.

Caitlin just has it. It's hard to describe her magnetism without sounding like a creepy magazine writer describing a young ingenue primarily by her looks, but Caitlin's Disney princess beauty, her long bouncy dark hair, and her big eyes are the first things you notice when you look at her feed. Her impeccable style, brightly lit photos, and picture-perfect life with her husband, Chris, and baby daughter, Kennedy, in North Carolina round out the enviable tableau.

Caitlin always shows up to our Zoom calls perfectly made up with amazing hair (the opposite of the typical lazy approach most of us have to our video call looks). On our first call, scheduled during a trip she had taken to upstate New York to shoot blog and Instagram photos in fall foliage, she wears a perfectly on-trend beanie topped with a huge fuzzy ball over her curls. In others, she is sitting at her desk in an office that looks like a West Elm catalog. She always looks so good, in fact, that I start to step up my game, hurriedly throwing on eyeliner and smoothing my mess of hair into a somewhat presentable bun or ponytail before turning on my camera.

I ask myself why I feel a little bit stressed about my appearance when we chat, or feel the need to look extra put-together. Caitlin's glowing and all-consuming perfection reminds me of what I am not: someone who is perfectly coiffed at all times. I don't generally have a problem with that, but I could see how someone could compare themselves to her and come out feeling worse.

It isn't just that she's traditionally attractive or has a nice life. Caitlin has a magnetism, a look, a vibe, that makes you want to be her friend and emulate her look. When you're picking out a dress for a first date, you want to wear one like hers. When you're bored on a fall weekend, you get inspired to round up your friends and head out to a farm to pose with apples in flannel, because when Caitlin does it, it looks so fun. When you're heavily pregnant, you want to pull up Caitlin's list of things she took to the hospital to deliver her first baby.

Women either want to learn how to be her, or they want to tear her down to make themselves feel better. It's a response we have cultivated since childhood. Caitlin is the most popular girl in school. It doesn't really matter if she was in real life, but at least online, she looks just as perfect as the girl in class who always had the right hair, the right backpack, and the right shoes. We all know that girl.

In real life, though, Caitlin doesn't give off the vibe of a Queen Bee. She is a bit reserved but open and kind, and she always seems happy to talk to me. She speaks in a soft voice and with a deliberate manner, often pausing to collect her thoughts before explaining herself. She smiles frequently and asks genuine questions of me and her managers at Digital Brand Architects, who always join us on our calls. She is thoughtful and purposeful.

The most passionate she becomes is when she talks about her

business, especially when she is defending it against those who may not take it seriously. In these moments, her voice rises to a higher octave, her words become sharper and more forceful. Caitlin knows her worth, and she wants to make sure everyone else does too.

"I work harder than, like, everyone I know," she said during one of our calls. "It did not happen overnight and it did not happen just happenstance."

The trouble is, her looks, her demeanor, and her industry have led some people to not take her seriously. Caitlin feels she doesn't get the respect she deserves. After all, she has more than a million followers on Instagram and has been blogging for a decade. She has her own clothing line with the online boutique Pink Lily, and she has worked with dozens of major brands. Blogging and influencing have made her rich, and she did it by herself.

The real world was different from what Caitlin had dreamed. When she began applying for jobs in New York, she got rejection after rejection. Finally, she got a job offer, from the same publication where she had completed the externship. The starting salary was twenty-seven thousand dollars.

Maybe, Caitlin thought, she could do it as they do in the movies—get ten roommates, or live on peanut butter and jelly sandwiches. There are girls, she reflects now, who would have done anything to make it work. (I would know. I did it myself. Although my strategy was fewer PB&Js—and more credit card debt, which I do not recommend.) But when she told her parents about the job offer and the twenty-seven-thousand-dollar salary, they balked. There was no way, they told her, that she would be able to live on so little. "You just can't do it," they told her.

So, she emailed the company back. Is there any way to get a higher starting salary? Nope, came the response from the HR rep.

"He was like, 'Well, this is what all of our employees make,'" she recalls. "And they do eat ramen noodles a lot, but they make it work."

Caitlin wasn't going to get this version of her dream. Even now, it still bugs her a bit. She says, again and again, it wasn't impossible. She could have done it, but it just would have been really hard. She sighs. Would it have been worth it? She'll always have a little bit of wistfulness about never living in New York and living that dream.

Ultimately, she listened to her parents. She turned the job down, and she didn't get another job offer at a New York magazine. She kept applying. The hunt took its toll, and she opened up to her readers.

"This summer I applied to dozens of jobs, wrote what felt like hundreds of cover letters, and went on several interviews," she lamented. "When I still didn't have a job, I was seriously starting to lose hope. . . . I began to think I would be stuck at home forever . . . and that something was wrong with me!"

It's funny because Caitlin is exactly the type of person you would expect could make it working at a magazine in New York City, but her dream seemed unattainable. If she couldn't make it, how could someone with less financial, racial, and social privilege ever expect to?

Instead, Caitlin accepted a job in public relations in Greenville, South Carolina. She wasn't making that much more money—just twenty-eight thousand—but it would be a lot easier to live on in the South.

"If you're waiting for something to happen, are frustrated with your life, or seeking change, have hope!" she writes. "God's plan for us is bigger and better than anything we could have imagined ourselves. His timing is always perfect. My adventure in the 'real world' is just beginning, and I'm so excited to have all of you along for the ride."

She never gave up her blog, though. Turns out, she was onto something.

"I always think about how I really, really wanted to work in the magazine industry and it didn't happen," she reflects. "I feel like I still do the same thing that I wanted, I originally wanted to do, like I'm an editor, I'm a stylist . . . But I created it for myself."

3

In even her wildest fantasies, Mirna could have never pictured this.

All around her were world-class athletes. None of them looked like her. They had gathered for a summit for ambassadors for the outdoor company Merrell, one of the most prestigious and well-known brands in hiking and trail running. Mirna stood among models and accomplished photographers. Two of the men had summited Mount Everest. Another had his own TV show. Only one other person was Black, and his résumé was formidable too.

Then, there was Mirna—the newest Merrell ambassador, feeling nervous and out of place. It was March 2016, and she had really just started her blogging career. For all her outward confidence, a feeling gnawed in her gut. What was she doing here? I mean, *look at these guys*. She wasn't a world-class athlete, a model, or a champion. She had certainly never climbed the tallest mountain in the world. Yet, here she was.

The feeling lingered. Later, she called her publicist, Margaux, from the hotel. Mirna hadn't really wanted to hire Margaux, not because she wasn't great, but because Mirna didn't believe a publicist would actually be necessary. But the interview requests via social media and email were a deluge that soon grew overwhelming. Finally, at the urging of a friend, Mirna found Margaux.

Shortly after hiring Margaux, Mirna got an offer. Merrell wanted to pay her a whopping forty thousand dollars to appear in a commercial, plus free stuff. The lump sum for just a week of work was nearly equal to Mirna's teaching salary for the entire year. Truthfully, she would have done it for free. She tried to not get *too* excited, not quite believing that anyone would actually pay her that much money for such a small amount of work. Ultimately, the deal fell through, as these things often do, but Merrell offered her a consolation prize: to become an ambassador. Mirna would get some free gear, get to travel a bit, and Merrell would comp her fees to some races. Mirna would have loved the forty grand, but this wasn't bad either.

But in her hotel room, Mirna unloaded to Margaux. "I don't think this is right," she worried. "This isn't my scene; these aren't my people. I don't—I can't belong here."

Margaux didn't coddle her. "You do belong there," she told her client. "These *are* your people. And you are going to see why."

Reflecting back on those feelings now, Mirna knows Margaux

was right. While no, she didn't look like the models, and it was true, she was no world champion, she offers Merrell, and the other companies she has worked with, a different kind of value. Because while she may have been different from all the other ambassadors in that room, there are more people like her in the world than there are people like them. Mirna could speak to those people. She *was* those people. She offered Merrell a chance to show that they care about and serve people like her: Black people, overweight people, or just anyone who doesn't look like an Olympic athlete. Those people are athletes too, and of course, they are potential Merrell customers.

For so long, Mirna didn't see herself in advertisements for running shoes, sportswear, or anything, really. If she saw anyone above a size 10 in *any* ad, they usually looked the same: hourglass figure, curves in all the "right" places.

But through her blog, Mirna was reaching people like her. Average people, not models or world-class athletes. People who exercise because they love it, not because they are punishing themselves or are trying to slim down before their wedding.

Now companies wanted access to this audience, and they were willing to pay Mirna for it.

"I'm the forgotten," she says. "And now, now here's our turn. Here's my chance to bring them into the fold."

During the 2010s, thousands of companies, big and small, had the same realization. Rather than the company hiring a famous, but not exactly relatable, actor or a generic model to tell people why they should buy their product, why not sell through a person consumers trust? Someone like Mirna, whom potential customers watch every day on their phones, speaks directly to them. The parasocial relationships formed when someone like Mirna lets followers into her life, they reasoned, are perhaps more powerful than even the most compelling commercial.

When Mirna tells her followers that she recently started wearing, let's say, a pair of trail-running shoes from Merrell that she loves—and here's why and oh, by the way, she can get you 10 percent off—it almost feels like a recommendation from a friend. The resulting effect from commodifying these parasocial relationships has been a boon both for companies and, of course, for people like Mirna, who is making more money than she ever has in her life. And it has created the influencer marketing industry, now valued at about $16 billion a year.

IN MY JUNIOR YEAR of college, I decided I needed to have a more exciting summer break. So, I applied to and was accepted into a summer program through my university giving students the chance to intern in New York City. The school was supposed to do the hard part of finding me an internship in my chosen field and arranging my housing. But when I arrived in the city, I quickly realized that my "internship" was actually a volunteer slot at a public access television station. It would have been fine, except for the fact that they needed me only a few hours a day, a few days a week. So, during my first couple of weeks, I spent a lot of time wandering down Broadway in SoHo, drenched in my own sweat, killing time window-shopping because I couldn't afford to buy anything, and waiting for my roommates to return from their big-city internships.

Luckily, another student in the program, Danielle, took pity on me and asked the bosses at her internship if they had room for one more (unpaid) intern. What did they have to lose? They said yes, and that's how I ended up working at a digital marketing firm called Talent Resources in the summer of 2010.

I had a lot of fun working at Talent Resources. One time, Snooki came into the office (in 2010 this was a *big deal*) wearing sunglasses so bedazzled she couldn't see out of them. One of the employees had a *real* American Express Black Card, which does *not* feel like a normal credit card. Dani and I went to a private performance at a trendy club by the teenage daughter of some big shot who was paying Michael Heller, the founder and CEO, to launch his daughter's pop music career. Kelly Bensimon, from *The Real Housewives of New York City*, was there too. We also helped two of the employees, who were trying to launch an alcoholic tea beverage business, host mixers a few times so they could hone their recipes. (That business eventually became Owl's Brew, a real, legit boozy tea brand now sold at Trader Joe's and Whole Foods that, in hindsight, was ahead of its time.)

Every day, Dani and I hopped on the 6 train to the office, an incredible brownstone in Murray Hill. We flopped ourselves onto beanbag chairs on the floor, where we posted up with our MacBooks, fiddling away until lunch, listening to the office gossip, and helping ourselves to free seltzer from the fridge. We bought bougie salads every day from the Guy & Gallard around the corner, piled high with all the fancy stuff we usually couldn't get (yes, I will take that avocado!), because our lunches were comped by the company. Then we would go home to our NYU dorm and party. We were living the life.

Michael was an entertainment lawyer and an Oz-like figure among the mostly female staff. I rarely saw him but he was often discussed. At the time I knew only a few key facts about him. He had made a name for his company, which I vaguely thought of as a "PR company," by working with Lindsay Lohan, and he seemed to have a lot of powerful connections.

I remember more about the parties and glamour of the job than what I actually did day-to-day as an intern. I had to kind of wing it because I was a journalism major and had really no idea what working in marketing actually entailed. But the one thing I do remember doing was actually core to Talent Resources' business and future success, and mine too.

I spent a lot of time thumbing through tabloid and gossip magazines to pull press clips for clients. But Talent Resources wanted something specific. What I was looking for was a celeb, usually a C- or D-lister, using a specific brand or product in a paparazzi photo. One time, I pulled a photo of Mario Lopez wearing a Tide detergent T-shirt at a gas station. Another time, I snipped a photo of Melissa Joan Hart using a certain brand of sippy cup while walking with her young sons.

Talent Resources was doing something novel for 2010: facilitating deals between these celebrities and brands. The celebrities would wear or use the product, get photographed, and then get paid for the exposure the brand got, as proven by the clips I pulled. The Talent Resources website credits Michael for being one of the first to do this, saying he "quickly realized that there weren't many active facilitators bridging the gap between brands and talent." I remember being shocked at how much Khloé Kardashian had made from one of these deals: tens of thousands of dollars for one photo.

Does this all sound familiar? What Talent Resources was doing back in 2010 was one of the earliest examples of sponsored content ("sponcon"), or influencer marketing. Celebs were getting fat payouts to integrate products into their daily lives.

What Michael and his team were doing was revolutionary. It laid the groundwork for the type of influencer marketing we see today. Famous people work with brands to integrate products into their usual content, expose the product to their audience, and get paid for it.

Indeed, Talent Resources is now known as a premier influencer marketing agency. I see their name and agents pop up all the time on social media. They have brokered deals between influencers and big national brands like Dunkin' and Fabletics, as well as Instagram-famous brands like Diff Eyewear and Teami Blends.

Nearly a decade after I interned for Michael, I caught up with him and his current business partner, Matthew Kirschner, whom I also worked with back in 2010. I wanted to learn about how they went from putting a Tide shirt on Mario Lopez to becoming one of the biggest players in the space.

It didn't happen overnight, by any means. At first, they told me, Michael and his team were seen as tacky, partnering with reality stars looking for a quick buck rather than cultivating relationships with "real" talent. Celebrities turned their noses up at hawking products on social media, and agents and managers didn't even want their clients on social media at all (it wasn't controlled enough, they thought). Even Talent Resources' competitors in the industry made fun of them for cashing in on "easy" deals with D-listers. Brands didn't want to partner with their thirsty talent either, not wanting to be associated with someone as lowbrow as a Kardashian. Besides, what good would paying for a post on social media do for them? They could see and touch a print ad or a commercial, so the return on investment was obvious. What if they didn't follow the celebrity? How would they even see the ad?

Despite all the skepticism, Michael knew he was onto something. One moment of those early days sticks out for him and Matthew.

Talent Resources is big into gifting suites at events like the Sundance Film Festival. Celebrities come, get their pictures taken, get a bunch of free stuff, beauty treatments, or food from companies that partner with Talent Resources, and leave. It's all very Hollywood.

That year, Moose Knuckles, a Canadian outerwear company, had sent products for the suite, jackets for celebrities to try on.

Instagram was barely even a thing at this point (this was around 2011 or 2012), but one person who was on it, and already had about a million followers, was Khloé Kardashian.

She came to the suite at Sundance and, after scoping out all the different goods, tried on the Moose Knuckles jacket. In a simple yet revolutionary move, she uploaded a photo of herself wearing it, tagging the Instagram handle of Moose Knuckles in the process.

Soon, Matthew was getting a call from people at the brand. *What just happened? What did you guys do? Our website just crashed.*

At that moment, Matthew recognized the power of this type of marketing, and what it had the potential to become.

But soon, Talent Resources wouldn't be brokering deals with just celebrities and reality stars. They would turn to influencers.

Caitlin's postgrad life seemed to be on the right track. She hadn't quite made it to New York, but she had at least gotten a change of scenery. After college, she moved to Greenville, South Carolina, a small but vibrant city she ended up falling in love with. She had met a boy, Chris, eighteen days after moving to the city, and she was excited about the relationship.

But once Caitlin got to Greenville, she didn't love her job. She hadn't studied public relations, but after giving up on her dream of working in magazine journalism, PR had seemed like a close enough alternative. The first few months were a challenge. She had to learn the business from scratch, she wasn't making that much money, and she didn't feel fulfilled. Her days were spent writing press releases for dermatologists and dental offices—not exactly the most exciting thing in the world.

Luckily, though, she had her blog.

Blogging was everything working in PR wasn't. She could be creative, write about fashion, and show off her fun side. Her real life provided her with a bunch of content ideas too, as she shopped for work outfits that were office-appropriate and furnished her first grown-up apartment. She couldn't wait to share it with her readers.

"Hello friends! I'm writing to you from my new apartment—yippee! I can't even tell you how much I LOVE having my own place. I moved in Friday, and even though I literally don't have any furniture for it yet, I'm so happy here. It's a bright little one-bedroom apartment with tons of character."

In the accompanying mirror selfie, Caitlin outstretches her arm to show off the furniture-less space. She's wearing a knit poncho and her arm is filled with bracelets, very 2012 chic (#armparty anyone?). She goes on, telling her followers that they will never guess where she found the perfect gold starburst mirror: Home Depot, of all places! And just thirty-five dollars, to boot. Could they believe it?

It wasn't easy balancing a full-time job with her blog, but Caitlin was determined to make it work. She made a content schedule for herself: She would post on Mondays, Wednesdays, and Fridays. She would spend her nights writing blog posts. She began posting her outfits and home inspo to new platforms, like Pinterest and Instagram.

The more Caitlin felt like the traditional working world was failing her, the more compelled she felt to work on her blog. She was a recent college graduate, and even if you're lucky enough to secure a stable job like Caitlin, entering the working world can be overwhelming and isolating. She was doing the worst and most menial work in the company for the lowest pay, sitting at the bottom of the totem pole. It could be stifling, especially when she wished she was doing something more creative. So, Caitlin found an outlet. The more frustrated she felt after a long day of work, the more fun blogging seemed in comparison. It was the bright spot of her days and weeks. Being a blogger was exhilarating.

Even though she was devoting time and creative energy to her hobby, Caitlin didn't really tell anyone she knew in real life what she was doing, besides her parents. She was embarrassed to tell her friends she was spending her time posting her outfits for strangers on the internet. Would they get it? Probably not. What if they thought she was weird? Her blog meant so much to her, and she was so excited about it, but she wasn't sure if people would understand it. The vulnerability of opening up her secret internet life to her real life felt daunting.

But as 2012 turned into 2013, a lot of other people, whom Caitlin hadn't met, had started to follow her. The more and more of her creative energy she poured into her blog, the more followers she seemed to attract. She became internet friends with other bloggers, who would link to and comment on her website. She would follow suit until she started getting so many comments that she couldn't keep track and comment on all their pages. Soon, she was getting about a hundred thousand page views a month. The more she posted and experimented with these new websites, Pinterest and Instagram, the more readers and followers she got.

Eventually, her cover was blown. Her college roommate had been browsing Pinterest, and there was Caitlin, in one of her outfit posts. "Caitlin!" her friend chided her good-naturedly. "We lived together in college—how could you not have told me about this blog?"

Brands began to notice her too. One day, Caitlin received an email from a representative for a company called Marleylilly. Marleylilly is the quintessential Southern shop, selling clothes and accessories, and jewelry with lots of bright colors and monograms. Like Caitlin's blog, they were based in South Carolina, and they were new—the company had started in 2010. Then, their big sellers were monogrammed necklaces.

Marleylilly told Caitlin they wanted to send her some of their products . . . for free. All she would have to do was post herself wearing them. Caitlin was gobsmacked. She couldn't believe this. All she wanted was to post about fashion, have fun, and put together outfits. Now she didn't even have to *pay for them*?

Soon another boutique followed, and another. Caitlin was in heaven. She had been posting her outfits just to to share with others, dutifully linking to every piece so her readers could buy it for themselves. She hadn't thought much about the fact that each time her readers clicked, she was helping the companies gain customers. But the small boutiques had, and they wanted Caitlin's followers.

Over the next few months, gifts from brands arrived more and more frequently. When she posted her free clothes on *Southern Curls & Pearls*, the boutique would sell out. Just like with Khloé Kardashian, Caitlin's tacit endorsement to her followers meant sales for brands. Her influence had made them money.

Caitlin would painstakingly link out every clothing item she posted on her blog as pure reader service. Her work was sending these companies a ton of new customers, and she got nothing. What's more, she didn't even realize she could get *anything* for her efforts.

In hindsight, she should have been charging the brands for this advertising beyond just free stuff, but that idea had barely even been tested yet, by anyone. Even when Talent Resources brokered deals in those days for what is now known as sponcon, they mostly were doing it through paid tweets rather than blog posts from noncelebrities. It's not that Caitlin was naive—it's that she was blazing a trail without a map.

In Caitlin's blogger life, the only people she could really rely on to bounce new ideas off of were other bloggers, who were participating in the trial and error of this new industry alongside her. In 2013, one of them, a blogger friend, sat her down.

"Caitlin," the friend told her. "What are you doing? You're selling out boutiques—you need to start charging for these services." The friend told her she could charge per post, and make herself a rate sheet. She could turn it into a whole big package: a blog post, a Pinterest photo, and an Instagram photo, all bundled together for a price.

Of course, she was right. Caitlin realized she had been underselling herself and her services. She needed a business plan. She had no idea where to start, but that was okay. She would figure it out. So, Caitlin did what many people do when they aren't sure how to summit a hill of adulting: she called her parents.

Caitlin and her dad sat down together and hammered out the details. Here's how much she would charge for a blog post, here's how much for Instagram, and here's a bundle together. One blog post and accompanying posts on social media cost eighty dollars. There weren't any guidelines for this kind of stuff, but it seemed fair enough. She sent her new rates to potential clients, and the brands accepted.

She was officially making money and attracting followers quickly. Caitlin joined Instagram, then a simple photo-editing app, in 2012. She began experimenting with the app, posting her "outfits of the day" or "OOTDs," and participating in trends like #armparty, in which you post a photo of your arm full of bracelets.

It was silly, but it worked. It was ridiculously easy to get followers on Instagram at the time. Just using a simple hashtag like #OOTD on your post would get hundreds of thousands of views (any wannabe influencer trying to do that now would barely get noticed). It was all about experimenting, seeing what worked and what didn't. Since influencing and blogging wasn't a job, there was no real pressure.

Offers for sponsorships also started rolling in for Caitlin. Swoozie's, a Southern gift store chain, paid her five hundred dollars to promote notebooks and tumblers they were selling. Many more boutiques began to reach out, offering her rates similar to Swoozie's. Eventually, she began to increase her rates.

By the beginning of 2014, Caitlin was making good money from her blog. In fact, her blog income surpassed her income from her day job. She began to wonder if she could quit and run *Southern Curls & Pearls* full-time.

The idea seemed absolutely crazy, but Caitlin was in the headspace to shake up her life a bit. At the end of 2013, she and Chris had broken up. The breakup sent Caitlin into an emotional tailspin. As she wrote on her blog, the subsequent months "were some of the hardest in my life."

"I thought my world had ended because I had lost 'the One'—the boy that I believed I would spend the rest of my life with," she wrote. (Spoiler alert: he *was* the One and they reconciled three months later.)

But at that point, Caitlin needed a change. She wanted to move home to Charlotte, away from the memories of Chris in Greenville and somewhere she would have support from her family to mend her broken heart.

So, she decided to do it. She took a huge leap of faith and quit her job to blog full-time. Was she crazy? She worried she was making a huge mistake. Would she really be able to support herself from this income for work that was barely even recognized as legitimate? She didn't know any other bloggers who were doing it full-time. What if she failed? Did she have enough belief in herself? As he always did, her dad encouraged her, pushing his daughter to take the leap. *You just have to take the risk*, he told her. *You just have to try it or else you won't know.*

Caitlin now believes her blog would have never become the success it did had she not broken up with Chris and, frankly, needed a distraction. She was so sad that she poured all her energy into her business, working a hundred hours a week. Although she had doubts, she said she was more excited than nervous. She prayed a lot about her decision. She felt strongly God was telling her that this was the thing she was being called to do. She had faith in her faith, and in herself.

That year soon brought another innovation that would forever change the way Caitlin ran her business. That new business was called RewardStyle.

FROM AN EARLY AGE, Amber Venz Box loved fashion. But, like Caitlin, she struggled to break into the traditional fashion scene. Amber claims to have been an introvert and quiet as a child, but now she is the epitome of a #bossbitch. With her fire-engine red hair and sharply applied makeup, Amber commands a room. She speaks at a fast but deliberate clip, pausing slightly every so often to ensure you are still following the narrative she is crafting for you. You immediately want to listen to what she has to say, and Amber always has a lot to say.

From a young age, Amber had a knack for entrepreneurship. In high school, she started making jewelry and selling it at local stores. As an undergrad at Southern Methodist University, she found internships in L.A. and New York, thinking she would be a stylist to the stars like Rachel Zoe, who'd worked with celebrities such as Lindsay Lohan and Jennifer Garner and was major at the time, with her own reality show. Turns out she hated being a stylist, so she tried out other parts of the industry through interning, eventually deciding she wanted to work at a fashion magazine.

After Amber graduated, she applied for jobs and internships, asking to be a closet intern or any lowly position. She never heard back from anyone.

"Oh well," she decided. "I will just make my own fashion business." Amber continued selling her jewelry and started working as a personal shopper. She went to a few local stores and offered them a deal: "If I bring in my rich clients and they buy things, you will pay me a commission on the items." Amber found a lot of success as a personal shopper but still wanted more. In 2010, she launched a blog, *VENZEDITS*, in order to build up her business, and hired a photographer to help her website look legit. On *VENZEDITS* Amber wrote about her personal style, linking to outfits she had worn, sharing sales and deals, and looks she coveted from designer runways. She didn't post a lot about her personal life or even write very much, choosing to focus on fashion and photos, but it was enough to get her featured in *The Dallas Morning News* as a local "blogger."

The label was funny to Amber because she didn't really consider that what she was doing was blogging. But she soon realized that this whole blogging thing was a huge untapped market. In 2011, she attended a blogger conference in New York, where she saw some big fashion bloggers, like Leandra Medine Cohen of the now-defunct Man Repeller, speak. At the conference, she began to ask other bloggers some of her burning questions. Mainly, how did they make money? After all, she was making a commission by doing the same sort of thing—marketing clothes to clients and getting them to buy them—offline.

"Well, I get free clothes," Amber recalled one blogger telling her. Okay, she thought to herself, so you live at home like me.

Amber thought that she could apply the same principles of her personal shopping business to what bloggers were already doing. Bloggers like Caitlin posted lists with links to everything they were buying, and presumably, people were clicking on these links and buying things. Shouldn't she, and all bloggers who were acting as personal shoppers, get a cut?

It was a revolutionary idea, and Amber was in the prime spot to test it out. Not only did she have her business experience, but she could also call on her boyfriend, Baxter (they are now married), to help with the technical side. They developed a simple prototype of what would eventually be called RewardStyle. (Amber picked the name. As she acknowledges, she is a very literal person.) RewardStyle partnered with retailers to make bloggers commissions. If the blogger linked out to the retailer using the RewardStyle link, they would make money every time one of their followers clicked on the link and bought something off the website.

Amber and Baxter officially founded RewardStyle in 2011. She spent hours on the phone over the next year, convincing bloggers to try out her new idea. She started by tapping her connections back at her alma mater, Southern Methodist University, where she knew journalism classes required students to have a blog. Some of them agreed to give her site a try.

She also trawled the internet for anyone she thought had potential. Early on, she reached out to a young woman who had been posting her fashion on Tumblr. She wasn't a big-name blogger, but Amber liked her style.

Amber tracked down her contact information. Once she got her on the phone and pitched her the idea, the blogger was reticent. "I just do this for fun," she told Amber, "and people aren't coming onto my site to buy things." Amber cajoled her. "Just hyperlink one item," she told her, "and see what happens. If it doesn't work out, we can drop it and just be blogger friends." Finally, the blogger did, linking out to a dress in a Friday post.

On Saturday morning, Amber woke up and checked to see how the link had performed. In less than twenty-four hours, the post had sold eighty dresses.

"And now," she tells me, "that person is one of the biggest influencers truly in the world."

Slowly, bigger and bigger names got involved, mostly through word of mouth. Turns out, Amber said with a laugh, everyone wanted to make money blogging. Getting the retailers on board was a little harder. Online retailer Shopbop was the first to agree to try it out, but it took her over a year to get another three stores on the platform. She spent all that time essentially clawing her way in, fighting just to get meetings with brands to hear out her pitch. But just because she got the meeting didn't mean she got a friendly reception. At one meeting with a big retailer, she was told that the company worked with magazines and models, not bloggers. "Candidly, I would get fired if I paid a blogger," she recalls being told.

If you've gotten this far, it shouldn't surprise you to learn that the biggest and most influential company in the influencer ecosystem was started not by a Silicon Valley entrepreneur eager to capitalize on a hot new market, but by a woman, herself an influencer, who realized there was a huge gap in the market that the "real" businesspeople were missing.

Since Amber founded RewardStyle and then its app, LIKE ToKNOW.It (now one company, LTK), its influencers have generated $9.5 billion in retail sales. As of 2021, the company said it was generating $3 billion annually in retail sales (in 2019, that figure was $1.5 billion). There are 130 women who have made more than $1 million from LTK, and companies have spent more than $1 billion on ad deals through the LTK brand platform. In November 2021, the company received a $300 million investment from the famed SoftBank Vision Fund, which valued it at $2 billion.

The numbers are staggering. Amber and her team are responsible for making 130 women *millionaires*. She single-handedly monetized platforms like Instagram, which until 2021 offered its users, who were creating content for it for free, no compensation for their work directly through the platform. Her company has physical offices on five continents, 350 team members, and more than eight million consumers who visit its platforms each month as of 2021.

Amber wants to know why she was the one who had to come up with a way to start paying influencers what they were worth to brands and social media platforms. Don't get her wrong—she is happy to have started her now-billion-dollar company. But she thinks she shouldn't have had to be the one to do it. The problem is, no one else was. The social media platforms, mainly Instagram and Pinterest, weren't sharing revenue with influencers, and more than a decade into the business, they still don't see influencers as an opportunity or support them as Amber knows they deserve. So she had to do something.

"I was on a mission to bring some legitimacy to these women," she told me.

ONE OF THOSE early adopters of RewardStyle was Caitlin, who, like many bloggers, first heard about it from a blogger friend. In Caitlin's case, that friend was Emily Gemma. Emily and Caitlin were quite a pair in those days, working together to improve their blogs. They would go on trips together, first going to an all-inclusive resort in Jamaica using a Groupon. They posted vacation photos at the same time and took pictures for each other to use in the future. Followers liked them together, and their friendship was both fun in real life and advantageous for their brands.

Emily offered to refer Caitlin to the service (another of Amber's ideas that led to a lot of its early growth), telling her she could actually make money doing things she had already been doing. Caitlin signed up but barely used it at first. She would make a few dollars here and there, but sponsored content was making her most of her money. That was about to change.

In 2014, shortly after she did the five-hundred-dollar deal with Swoozie's, Caitlin began to take RewardStyle more seriously as she transitioned to blogging as her full-time job. As she used it more consistently, her profits began to grow. The big turning point for her use of the platform, though, and for her business overall, was when she decided to attend one of RewardStyle's first annual conferences, in 2014.

The conference was pretty chic—especially for 2014, before big splashy influencer retreats like those held nowadays by Amazon and Revolve. Held over two days in Dallas, the conference, according to a *Texas Monthly* report, featured parties with brands like Kate Spade, one that got in early with influencers. The magazine reported that hundreds of bloggers attended, some of whom were raking in anywhere from twenty thousand to eighty thousand dollars a month from blogging. "As the party got under way, the bloggers, uncased iPhones at the ready, started snapping selfies. A gaggle of them, in Kate Spade shorts and party dresses, posed for a photo that Kate Spade itself would later Instagram," the magazine noted.

The conference was a revelation for Caitlin. It was the first time she had ever met other women who were blogging full-time as a career and making a lot of money doing it. Big-name bloggers like Aimee Song of *Song of Style* wowed her, and she felt incredibly motivated and inspired. She also learned more about the platform and was taught how she could best utilize it.

It also marked the first time Caitlin realized just how much money some people were making blogging, and how much potential her business truly had. One of the blogger friends she had made with a similarly sized following told her she had gone viral on Pinterest earlier that month by posting a photo of herself showing off her outfit. The photo drove a ton of traffic to her blog and led to a ton of clicking on her RewardStyle link. She made seven thousand dollars just from that one outfit.

Caitlin was blown away. She had no idea that kind of money was even possible from blogging. That much money, from one day's work, was more than she had made in two months at her old PR job. *I want to do that*, she thought.

By November 2014, she had made RewardStyle's list of the top ten highest earners on the platform. Soon, the money she made was truly mind-boggling. One month, she made three thousand dollars. The next month she doubled that. Then she doubled *that* figure. Then suddenly, she was pulling in thirty thousand dollars a month in commission. And it kept going up. There didn't seem to be a ceiling on how much money she could make.

"It was such a roller coaster, just thinking . . . the sky's the limit, really," she said.

IF YOU SPEAK to any influencer about these early days of affiliate linking and sponcon, many of them use similar terms to describe it. A deluge. Overwhelming interest. Money, and lots of it. Absolutely zero regulation, standards, or uniformity across posts or influencers. Basically, it was the Wild West.

Mirna laughs when she thinks about her early days in the business, trying to navigate this role as an influencer, a term she doesn't love but will apply to herself in this case. She had no clue what in the world she was doing. After her features in *The Wall Street Journal* and *Runner's World*, she and then Margaux fielded countless requests for interviews, writing gigs, sponcon, and ambassadorships.

The opportunities just kept falling into her lap. Mirna hasn't ever marketed herself or sought out opportunities to make money as an influencer (though she does market herself as a speaker). People have always just come to her. In the beginning, it was pretty overwhelming stuff for a teacher to try and navigate on her own.

"The requests were just coming in, and that was why I had to get a publicist because we couldn't keep up," she said. "I couldn't keep up with the demand. It was a very cool problem to have."

Still, it was all a little crazy. At first, she said yes to every opportunity that came her way, because how could she not? But of course, that couldn't last. After all, she was a full-time teacher at a boarding school, where she was also coaching sports and running a dorm. It was all a little much.

"It was definitely very time-consuming and overwhelming and tiring. But I was so full of energy because I was like, this is so cool," she said. "Who does this happen to? This is amazing. And so I just tried to say yes to everything, which, you know, I'm glad I did. I can't do that anymore."

Mirna had been plunged into the very nascent industry of influencer marketing, and had to teach herself to swim. The questions these companies asked her made her head spin. They asked her about assets, deliverables, and ROI. Mirna was like, "By assets, do you mean photos? Okay, yeah, I can shoot you some photos. You want to pay me how much money? Sure, I guess that works."

To make this work, Mirna had to rely on herself, and a small group of friends she knew who were also trying their hands at influencing. It's not like she could sign up for Influencer University. There is no such thing as a national association of influencers telling her how much to charge and how much to work. For Mirna, the whole thing—this whole new, strange career—has been trial and error. She is completely self-taught, learning photography tricks through examining other influencers' posts and poring over engagement stats to figure out what people wanted to see.

The first "influencer" deal she ever did was with REI, the sporting goods store. They approached her, asked her to do two or three posts on social media, and offered her five thousand dollars. She didn't negotiate or anything; she just took the deal. She already had a job, after all, so it's not like she needed the money. And they were just asking her to take some photos from her everyday life. How hard could it be?

Then, more companies began to ask her to create social media content for them. Around 2017, she decided her base price would be one thousand dollars, which she remembers coming up with after speaking with other people she knew who were influencers. But again, she didn't have any references to consult, and she was mostly flying blind. All of them were.

Over time, Mirna began to educate herself. She began to really consider her value to these companies, and study what it truly meant to be a creator. She learned about the value of her engagement rate and decided what she was and wasn't willing to post. As she has learned and become a veteran, she has always tried to pay it forward to the other influencers who frequently ask her for advice. "I think there's no precedent for that," she says, "and so I also don't want people to get fucked over."

To Mirna, sponcon was and is a means to an end. By being an influencer, she could spread her message of racial and body diversity to potentially millions of people. She could reach not only her community, but communities that may never have heard from someone like her. That was a power that she had never really dreamed of, but she was eager to take it.

4

My first impression when I entered the home of Shannon and Dallin Bird was how clean it was.

To some reading this, this may come as a surprise. After all, Shannon's reputation on the internet is that of the zany, messy mom. But the reality is different.

Shannon hates clutter. When I visited, she was in the middle of redoing her primary bedroom, and looking at the mess, the torn-up floors, the contents of her closet strewn about to make way for the workers, seemed to physically repulse her. All she wants is for everything to be in its place, to feel a sense of control over her surroundings.

For the past decade, though, Shannon's life has been anything but orderly. She never expected to spend her twenties having five kids in rapid succession. But she did, one after another, without any time to catch her breath between toddlerhood and late nights breastfeeding a newborn. She never expected to become a mommy blogger, to become internet famous and showered with gifts from brands that, while nice, sometimes felt suffocating. And she certainly never expected to spend an entire decade of her children's childhoods being watched by a peanut gallery online, to have every decision she made and mishap and family drama dissected by those who felt to her like they were watching her every move, breathing down her neck.

Shannon had always dreamed of other things. Of escaping picture-perfect Utah and the judgment she has always felt there, and making her life on her own terms. So far, though, it hasn't turned out exactly as she had expected.

Shannon grew up in a suburb near the one where she and her family currently live, one of several middle children in a family of nine. According to her, they were "dirt poor." Her father struggled to maintain a job and the family finances were tight. Among the Mormon community, Shannon never felt like she fit. Because her family didn't have much money, she couldn't compete with the wealth often lavished on her classmates by their successful parents. She also never seemed to do things exactly right.

Shannon's family had come from Georgia, and though they are Mormon, they weren't used to the unspoken rules about how young girls were supposed to dress and behave in ultraconservative Utah. Her clothes were always a bit too skimpy, and her swimsuits too revealing. Her mom would get calls from other moms at the pool that Shannon's bathing suit showed too much skin. Shannon worried that they were calling her a bad influence behind her back. She always felt singled out as a "bad" one, a messy one, someone who just didn't fit into the perfect Mormon girl ideal.

By the time Shannon graduated from high school, she was ready to make a change for herself. She worked hard and went to college. She vowed not to date any men from Utah, in hopes of getting out of the state and out of the underdog role she felt she had been cast in for good.

So how did Shannon end up living smack in the center of the Mormon influencer heartland, regarded as the epitome of Mormon mommy bloggers, and discussed as one of the most controversial influencers on the internet? She wonders the same thing. It's almost like her life won't stop repeating itself, over and over. She used to get whispered about behind her back for her "skimpy" swimsuits; now she worries the moms at her children's schools are judging what she posts online.

Shannon worries that this dynamic is hurting her kids socially. Her oldest daughter, Holland, has complained that girls in school have asked her if she really did things like cut her own hair, after, presumably, their moms saw an Instagram Story of Shannon's and told their kids. They are judging the character she plays on the internet, which I am beginning to think isn't the totality of who she really is, despite the impression of complete, unguarded authenticity she creates.

Where did that character come from? It evolved slowly, over time.

When Shannon first met Dallin at a pool party, she was thrilled (on her blog, she wrote her first impression was that he was "hott"). He is a decade older and from the other side of the country—the complete opposite, she thought, of any of the boys from Utah. However, Dallin loves Utah, and after they got married and had their first child, Hudson, the Birds stayed. Shannon had wanted to do something different with her life, but she soon found herself as a stay-at-home mom. She didn't need to work to make ends meet, but she was craving an outlet that was all her own.

She started blogging, at first because it was just what everyone else in her community was doing. Growing up, Shannon was friendly with a family named the Skallas, whom we met in chapter one. They had four daughters, two of whom, Rachel and Emily, were around Shannon's age. Over the years, the women kept in touch and ended up settling in the same community after they got married. Rachel and Emily both had started blogs. Rachel, now Rachel Parcell, called hers *Pink Peonies* and Emily, now Emily Jackson, called hers *Ivory Lane*. Both were finding a ton of success in the industry, getting free clothes and sponsorships. They encouraged Shannon to try it out as well.

So, Shannon started a mommy blog and an Instagram. It was 2013, and life was good. Hudson was almost a year old and Shannon was a stay-at-home mom. She had snagged Dallin, who had been, by her own estimation, "Utah's most eligible Bachelor." She had a house and a lot of cute clothes and purses, and had just gotten a boob job.

Shannon was proud of herself, and she wanted her family and friends all over the country to be able to easily read about her "life updates," as she called them. (Dallin is one of eleven kids and Shannon is one of nine, so keeping everyone in the loop was hard.) So, she set up a blog, which she named *Hudson a la Mode*. The name, she wrote, stemmed from her love of ice cream and a shop she and Dallin had visited frequently early in their courtship. "I want all my readers to have an à la mode experience when they visit my blog," she wrote. "Like what ice cream does to me."

Then, the brands came calling.

The first were small businesses—Etsy shops and the like—who began to message her. One of the first Shannon can remember came in the early days of her blog. Halloween was coming up, and a shop owner reached out to ask if they could send Hudson a free costume. If so, would Shannon shout them out on her blog?

When Shannon realized strangers were reading about her adventures with Hudson, she was stunned. She thought only family members were reading, but apparently, her blog had been found by other people (many readers found new blogs in those days through online tools that would surface similar content, as well as links from other bloggers). But once Shannon and Dallin realized that her blog had moneymaking potential, they got to work setting up marketing tools like Google Analytics to track her audience and work on growing her blog.

Next came the free stuff.

The first packages began to show up soon after she signed up for RewardStyle, and they just never stopped. Kids' clothes. Toys.

Clothes for herself. Four hundred dollars' worth of swimsuits for kids, every year. Expensive kids' shoes that retail for eighty dollars a pop. Free hotels at Disney World, where her kids had the run of the park like they had won some sort of contest. A trip to a beach house in North Carolina, stocked with food.

When Shannon was pregnant with one of her kids, eight new car seats, from different brands, arrived at her house. When she traveled, like to Vegas or New York Fashion Week, she'd get DMs asking where she was staying. Suddenly, boxes and boxes of stuff would arrive at her hotel room. Her youngest daughter London's room was fully furnished for free. Shannon picked everything out and then posed for photos when it all arrived on big freight trucks.

The family started to make jokes about it. For Mother's Day one year, Dallin and the kids wrapped a bunch of things that PR firms had sent to Shannon that week and gave them to her.

Shannon thought it was funny. She almost posted about it on Instagram but didn't think it would go over well. People would probably just think she was bragging.

At first, it was cool. Shannon never had to buy her kids clothes, really, or clothes for herself. Her kids got brand-new swimsuits and shoes every year from high-end children's brands like Janie and Jack, significantly decreasing the number of things she needed to buy for her growing brood.

"I was so appreciative because those suits were forty dollars apiece and I got new suits every spring and winter," she said. "And those shoes, each of them is eighty bucks."

Shannon had a genie in a bottle. If she were to post tomorrow that she really needed a stroller, her inbox would instantly be filled with brands offering to send her a free one. One of the coolest things that happened in her early blogging career was getting a box from Kate Spade. She opened it up; a cute purse and a note were inside. When she followed the note's instructions and logged on to a website, it showed her pages of Kate Spade bags and told her to pick five.

"I was having a panic attack" from excitement, she said.

The best product she ever received, though, was free light fixtures for her home. The company contacted her and asked if they could send her enough to furnish the entire place, whatever she wanted. She ordered chandeliers and pendants, eight in all, and it took semitrucks to drop them all off. She estimates the total cost at twenty thousand dollars.

The opulence that Shannon was being showered with every day by brands raised some uncomfortable questions. Why her? And why were so many influencers like her—young, beautiful, Mormon moms from the same area of Utah?

It may be rather simple. I asked the Birds, and to them, the answer to the "Mormon mommy blogger" question is rather obvious. As Dallin explained while Shannon nodded along, the phenomenon came about because it is the perfect marriage of two factors: brands looking for the hottest, most desirable spokespeople, and young, ambitious women caught between two worlds.

It was an interesting time to be a young Mormon woman.

Shannon, Caitlin, and I are about the same age, and we grew up surrounded by the girl power mantras of the nineties. As a millennial girl, I was surrounded by women who worked, and was constantly told by my own mother and by society that I could "do anything a boy could do."

I remember as an elementary school student truly believing nothing could stand in my way. I could be whatever I wanted—a doctor, a lawyer, a soldier, or a construction worker. The reality was much more complicated, of course, but that didn't sink in until later.

Shannon and all the other Mormon girls she grew up with were surrounded by these same "girl power" ideals. Many of them were extremely driven, attended college, and had ambitions and career goals. However, they also belong to a culture in which it is expected they will settle down at a younger age than average, and have children earlier than the typical millennial woman.

As Shannon and Dallin, who was also raised Mormon, explained to me, this isn't something they are forced into. It is just a fact of their culture. Once their children are born, many of these young ambitious women choose to stay home with their children or work part-time. It was expected by their families and peers, even when the popular sentiments of their time seem to be telling them the opposite.

"It was really hard for my parents to even grasp why I was even going to college," Shannon said. "They're like . . . , 'You're never going to use a degree, why are you going to school?' And they didn't help me at all. They thought it was, like, a waste of time and money."

But these women wanted more. They were going to abide by the expectations of their culture, but they were going to do it their way, and that way was by blogging.

"So you've got young, attractive, pretty well-to-do, college-educated girls that go on the internet and start seeing blogs and reading magazines like *Cosmo* and *Glamour* and *InStyle* and *Vogue*, right?" Dallin said. "So they started journaling online a little bit."

Couple that, Dallin said, with the fact that Utah has always had a bent toward internet and technology businesses and that Mormons are encouraged by the church to journal from a young age, and the industry was born. Once the bloggers started to realize that their online diaries could be more than just a fun hobby, they were able to use their education and drive to leverage them into bigger and bigger opportunities.

Once one woman started doing it, it began to spread. They would see women like Rachel Parcell and Amber Fillerup Clark of *Barefoot Blonde* creating these careers for themselves, and these were just average women in their own neighborhoods. They didn't have any big Hollywood connections, or a leg up. It was attainable.

"Other girls were like, 'I can do that,'" Dallin said. "'I know them, this isn't untouchable. I could start a blog and create a little website and do stuff.'"

He added, "It's really because they're really smart, aggressive women who do have kids who are like, 'I don't want to go home and do nothing.'"

Dallin compared their area of Utah to another hub of tech innovation.

"It's like, why are there so many different tech companies in Silicon Valley?" he said. "Because one started, guys left [for a different company then] started another one . . . [and] this ecosystem is created. So there's an ecosystem that started ten years ago in Utah with bloggers."

These Mormon bloggers, though, did have a bit of a leg up. They were like catnip for brands, especially high-end brands that catered to mothers.

Consider Shannon and her peers from the perspective of a brand. Shannon is young and beautiful, with blond hair, big eyes, and a penchant for posing. She is always dressed to the nines, with her hair curled and mascara firmly in place. She also has a bunch of kids. If you're a brand, who could be a better model?

It may be outdated and sexist, but many companies still want their clothing or accessories modeled by a mother who fits a certain image: hot, stylish, and young. This was even more true in 2010.

This glam hot-mom persona was also cultivated by the women themselves. Remember, these are women in their twenties in notoriously image-obsessed Utah. And especially in the beginning, when radiating perfection, rather than projecting "realness" like modern-day influencers, was a requirement, this image was even more powerful.

Shannon believes that some of her success as a blogger is because of the type of mom she was in her twenties, compared to the more laid-back one she has become in her thirties. When she was young, she was much more glam and label-obsessed, always wanting to have the best and most fashionable thing. She credits this to her husband, who not only has a very bougie sense of personal style but was older than she was, so he had the money to invest in his own wardrobe. For Shannon, who had always been frugal, her husband's tastes were a whole new world.

"I went from, like, not even shopping at all to, like, Neiman Marcus for my baby stuff," she said. She would dress Hudson head-to-toe from Neiman's, even in baby Gucci. She was one of the first people she knew to buy the then-ultra-trendy Bugaboo Donkey stroller, which retails for nearly two thousand dollars. Shannon had the double, even though she had only one kid at the time.

For people her age—then twenty-one—she says, "that was, I guess, as much as someone's car."

So a baby clothing brand sees someone like Shannon, with her Gucci-clad baby sitting in his two-thousand-dollar stroller, and thinks "Bingo. This is the type of mom who would be perfect to send our luxury baby shoes, our clothes, or our Montessori-approved toys." And then Shannon gets it all for free, she posts it, and they get customers who are drawn in by her posts about her fashionable, fabulous life as a young mother. Everybody wins.

It's hard to overstate how lucrative the relationship between mommy bloggers and trendy baby brands has been for both parties. In 2021, I wrote a story for *BuzzFeed News* in which I examined the rise of a handful of these fashionable brands and the moms who said they were influenced to buy them over Instagram. In that story, I cited a December 2020 report from Fortune Business Insights, which claimed that these relationships were fundamentally changing the baby apparel industry. They cited "growing social media influence and changing fashion trends" as one of the key factors driving the trend among modern moms.

Of course, it also has been lucrative for the Birds. Shannon doesn't just get free stuff, she also gets paid for sponsorships she does with brands. In the late 2010s, she was making about sixteen thousand dollars a month from sponsorships. Shannon called it her "squirrel cash," a way not only to support the family finances but also to feel independent and buy things for herself. She even was able to support the family for a while after Dallin quit his family's business and ventured out on his own. "It's this drive I've always had because I've always had to work for whatever I had," she said. "I like to have a little bit of my own money."

Once Shannon had this career, she then had to consider who exactly she wanted to be on the internet. She soon realized she had no interest in being the perfect Instagram mommy, preferring to be "Adam Sandler."

As I got to know Shannon, I began to realize that her unique influencer career is deeper than it seems on the surface. Online, she can seem like kind of, well, a mess. Her kids are screaming, her minivan is dirty, and she's constantly making faux pas, like posting a video of her youngest daughter, London, sucking on a tube of lip gloss like a lollipop or misspelling words on Instagram Stories. (Shannon acknowledges she isn't a great speller but tells me that has nothing to do with her intelligence.)

To her online critics, Shannon's life is a dumpster fire they can't look away from. One of the first commenters on her thread on the snark website Get off My Internets called her "one of the dumbest bloggers I have yet to come across." The criticism is unchanging and unrelenting, rarely abating for more than a few days.

"Some people just shouldn't be allowed to procreate . . . or talk on Instagram. She's one of them," wrote one commenter on a post making fun of her.

In real life, though, the Birds seem pretty normal. Is Shannon's online persona, then, just a character she is playing? When I asked her this, at first she said no, she's not making stuff up for the internet. But then Dallin interjected, yes, she is playing up her wacky side for the internet. Shannon, he says, is much more than what she shows her followers.

"Okay, maybe yes," she then said. "It's all real stuff that happens, but I just show the funny parts."

But also, it's possible that at least in the beginning, she *couldn't* be the perfect Instagram mommy. Those women were so much more serene, so much more put-together than she was. Their kids never seemed to run and scream; they were content to play quietly with aesthetically pleasing wooden toys and smile happily for photo after photo. The perfect mommies never seemed to get overwhelmed, they never raised their voices at their kids or got mad at their husbands. They had time, somehow, to nurse their infants, cuddle their toddlers, and bake sourdough bread from scratch. They could have all-white furniture, because who could ruin it?

That just wasn't Shannon's life. If sometimes it seemed chaotic on the internet, well, maybe that's because it was. Sure, a lot of people in her church had a lot of kids, but Shannon had had a lot of kids very quickly. She was only in her early twenties when she had Hudson, and the next ten years passed by in a blur in which she was constantly either pregnant or breastfeeding. She struggled to have any sort of identity outside of being a mother, and often she couldn't manage everything.

So yeah, sometimes her house and her car were messy and her kids did look a little disheveled. As someone who describes herself as type A ("People online would never believe it, but I am," she says), Shannon experienced a ton of anxiety during those years when her kids were young. Sometimes she was crawling out of her skin because of how little control she felt she had over the constant demands of raising so many young children. She never had free time, describing herself as an "octopus," her two arms flailing trying to do the job of eight.

Posting on Instagram about it, though, actually made her feel better. If she could make it seem on the internet like she thought it was hilarious that her kids were tearing her house apart, maybe she could start to believe it too.

"Posting a video or whatever is just trying to make it funny," she said. "I'll look back and think this is funny, but I'm drowning."

Shannon's Instagram feed from when her older kids were younger is actually not as chaotic as she remembers, but it also isn't as perfect-looking as other big accounts. But she definitely kept it real, and her feed has always looked more like that of an average mom in the trenches than a serenely perfect Instagram mommy. Her kids sometimes aren't looking at the camera or are making silly faces. Their hair or house doesn't always look perfect. But in many photos, both she and her children look happy and fulfilled.

While Rachel's and Emily's daughters always look like living dolls for dance recitals on their Instagram accounts, Shannon regales her followers with a tale about how Holland thought her recital costume was a swimsuit. So, right before the family piled into the car, Holland ran to the neighbor's house and happily got soaked with a bucket of water.

"Although she was the only girl in a wet costume with no curls and makeup, she danced her heart out and made me so proud," Shannon wrote on Instagram. Her candor and ability to keep it real won her some genuine fans. Over time, Shannon began to feel a sense of community from her followers as well. Maybe if they could relate to what she was feeling, things weren't as overwhelming as they seemed. The more she shared about how she was really struggling, the more she felt a little less alone in what she deemed her "drowning state."

Plus, as a stay-at-home mom with a ton of babies, it was nice to have people to talk to.

"It's, like, an easy way to get friends because I can't find a babysitter with five kids," she explained. "So I can sit on my phone and talk to people. It's, like, a social outlet."

So, over time, her persona evolved. She began to downplay her type A side and play up the "crazy Birds" side of her life. She shared the video of London sucking on lip gloss. She snapped a photo of one of her kids playing in the dryer. She joked, "whose kid is this," on a photo of one of her boys sneaking a piece of candy out of a bin at a store.

This persona did win Shannon some fans. But it also made her into an obsession with another group of people: the haters.

Almost as soon as Shannon started blogging regularly, people began to make fun of her on Get off My Internets, or GOMI.

At first, Shannon was mocked only as a side character on threads dedicated to snarking on other bloggers, like Rachel Parcell. In July 2013, Shannon was deemed snarkable enough for her own dedicated thread.

"Ok, so based on comments in the Pink Peonies, Ivory Lane, and Worst Fashion Blogging Threads, I thought I would start a thread for this little gem!" wrote the user who started it.

While Rachel's and Emily's haters tended to hate on the sisters for being too perfect, Shannon was mocked for the opposite. From the beginning, a narrative emerged about Shannon centering on two key judgments of her. One, she was not very smart—"Bird-brained," as many joked.

"Bird a la Mode is one of the dumbest bloggers I have yet to come across!" said one anonymous poster in one of the first comments on Shannon's thread. "I can't believe there are so many attractive ex dancer mormon girls landing rich husbands who all happen to blog about their lives. Really blows my mind!"

The second pervasive narrative about Shannon is that she is, as she says in her own words, "a shit mom." The first time she noticed this narrative was when she posted about having a miscarriage in 2014, in between her second and third children.

Now, Shannon said, women are praised for opening up about hard things like miscarrying. But back when she did it, she got a ton of backlash. People accused her of just looking for attention, about not really caring about the baby she lost.

"I read, 'You're an attention whore,'" she recalled. "Now it's like, 'Oh, let's be authentic, let's be real.' But I'm like, 'Yo, when I first was real, people would murder me for it.'"

Since then, Shannon has been the subject of almost daily posts on GOMI, and the vitriol shows no sign of slowing down.

It would be impossible to write a book about influencers without mentioning GOMI. This internet chat website is full of hundreds of forums, each about a different blogger, influencer, or another type of internet personality. It is an incredibly active forum and has not ceased chattering for the ten years I've been aware of it. The discussions range from the juicy (so-and-so is getting divorced!) to the mundane (I hate her outfit!). It has its own subculture, slang, and emojis. Often it's the only place you can get unfiltered gossip the mainstream media can't, or won't, report on (because they would get sued). The website also is one of the only places where you can discuss online personalities outside their own platforms. Picture the entirety of *Us Weekly* or *Entertainment Tonight*, but all the chatter is uploaded to an anonymous forum that looks like it hasn't been updated since the midaughts.

GOMI is not a very nice place. The commenters are often nasty to both the bloggers they follow and each other, and the speculation can get dark and rather twisted. Women are called fat and children are called ugly. Page after page of comments will speculate on whether a blogger's child has delays, either physical or mental, by examining their photos and videos. GOMI commenters have accused a blogger's husband of faking cancer (he later died). They have, according to a few bloggers, tried to track them down in real life or sent threatening messages to their homes. As the years have gone on, the website has only become more cruel, full of racism and sexism, and perhaps inevitably, it has become a pro-Trump cesspool in certain forums. These days, it is known as one of the last remaining places on the internet where you can be openly racist about Meghan Markle and be celebrated for it.

GOMI was started by a woman named Alice Wright around 2008, according to a 2016 story by *The Guardian*. Over the years, Alice, a Gen-X computer programmer who sometimes goes by "Party Pants" on the website, has become almost as fascinating a subject as the bloggers her site mocks, with people on other snark forums, like Kiwi Farms and the subreddit Blogsnark, dissecting the publicly available information about Alice.

Her past is shrouded in mystery (at least nothing I can report), and no one is really sure why or how she came to become the webmistress of a blogger snark forum. Here are a few things about her that I and others have gleaned from her public Instagram profile and posting on GOMI: she likes cats but also has had dogs, she lives on the Upper East Side of Manhattan (but used to live in Brooklyn), and she frequently attends galas and charity events in big ball gowns. She seems, according to her posts about "crotch fruit," to dislike children. She is a polarizing figure, who has shape-shifted into multiple different personalities over the decade-plus her site has been active. I have tried to interview her for many years, but have never been able to get a response from her. She's also blocked me on every platform, so there's that.

Alice has long claimed in interviews to be a computer programmer, but GOMI is a clunky website and has gotten only harder to use over the years, laden with banner ads that make it lag and freeze. It's also unclear if running GOMI is her sole source of income, but she has frequently asked community members for donations over the years to keep the site running. I recall vividly a campaign when I first started reading GOMI, in which Alice claimed she would have to imminently shut the website down, only to be saved by a flood of donations from good-hearted snarker patrons.

In a 2012 interview with the Daily Dot, which dubbed her possibly "the most reviled woman on the Web," Alice claimed the website barely broke even and described her management of the site as an act of service. She told the news outlet that she believes GOMI is doing good in the world. People want to be able to call out bloggers for their bad behavior, she said, and she's providing that.

"It's a relief valve," she said. "In my opinion, I think having a place to get things off your chest is healthy. My goal is really just to give people a place to say what they feel like they can't say to some of these bloggers. A place where they can give their opinion without being shut out."

One GOMI commenter, though, put it more bluntly in the *Guardian* piece.

"It is pretty damn fun sometimes to make fun of these idiots. Sometimes they are just so dumb in the things they do you can't help but to laugh," they said.

I've always been fascinated by the people who post on these forums, especially the ones who do so prolifically. I would categorize myself as an internet lurker. I read several forums on Reddit religiously, as well as Facebook Groups, and for many years would read GOMI consistently as well. However, I have never posted on GOMI and very rarely do on other forums. I guess I like to observe conversations, but I don't want to contribute myself.

Part of me also doesn't want to cross that line on snark forums. Are there people who annoy me on the internet? Sure, but

I have never felt comfortable making fun of someone in that way, even if I did so anonymously. I don't know if that makes me more morally sound in some way—after all, I have *read* some horrible things on GOMI—but I don't want to fully engage or cosign it. Especially because I think that 95 percent of the snark I see is rather petty, and for the most part is more needlessly cruel than it has to be.

I wanted to hear from those who actually do post, though, and see why they do it. I figured I would get the most honest responses if they were anonymous, so I used an app on Instagram to allow people to send in why they participate in GOMI and Blogsnark without having to share their names. I got a deluge of responses, which I was kind of surprised about. It seems people wanted to get their reasoning off their chests.

A lot of the people who responded had a similar trajectory for finding the sites that I did. They followed bloggers and obsessed over them, but had nowhere to discuss them. GOMI was refreshing. It was the only place you could gossip about drama in the space, or call out people you thought deserved it.

"I've been on GOMI since 2012 or 2013," one person said. "I'm not a troll, I'm someone who is genuinely interested in the industry and always curious about people . . . I don't enjoy or partake in the ruthless takedowns or irrational mockery or complaints. Some people on there are just hateful and it's not interesting or funny. But there's also some insightful, responsible critique that influencers should note. Also, sometimes you just need to vent about the cringey/absurd things influencers do."

"Nobody in my life gives a shit about influencers," another wrote. "I don't love the overly negative nature but I'm truly just in it to discuss an interest."

Gossip, after all, is a fundamental part of human nature. However, many people agreed that while GOMI used to be a place for more fun gossip and critique, it has slowly devolved over the years into a place of cruelty and wild speculation. More than one called it "hateful" and a "cesspool," describing some of the rhetoric as "genuinely cruel" (like when, for example, people call an influencer's child ugly). Some cited an incident we've already discussed as their breaking point, in which users accused a blogger's husband, a father of five, of faking cancer. The amount of speculation and conspiratorial narratives like this one was enough to make many posters squirm.

"At first I enjoyed having people to talk about crazy bloggers with, but eventually I just felt icky about it . . . so many people making things up whole cloth just makes me so uneasy," one person told me.

In fact, the Reddit forum Blogsnark was started in 2015 because people were fed up with the toxicity of GOMI, and many people who wrote to me said they had also made the switch because it made them feel less gross to consume.

It's a fine line, though, between harmless chatter and OMGs, and nitpicking and cruelty. Many people expressed they are now over Blogsnark too, because it just seemed to be entering the same cycle (others complained that the forum was actually *too* nice because it is heavily moderated). Several people told me that they broke the snarking habit and found they felt better after.

"I used to read GOMI pretty frequently," one told me. "I realized it was making me negative about everything. I couldn't see someone happy on Instagram without thinking something negative. I haven't been on the website in months and feel like my relationship with social media is much healthier."

The proliferation of the forums is bad for the mental health of influencers as well. In my influencer survey, many of the women said the stress of dealing with online harassment and criticism was one of the most difficult parts of the job.

"The hardest parts of being an influencer for me are maintaining the privacy and dealing with being spoken to or about as if you aren't a human being," one said. "Whether it's DMs or gossip forums, it's awful to have people talking trash about you (usually false) and often not really be able to defend yourself."

Another woman described the criticism as "constant," and others said it always weighed them down. The stress of never knowing if they were doing or saying the right thing, and many times being picked apart for it anyway, took a toll on their mental health.

PERHAPS SURPRISINGLY, SHANNON actually enjoys parts of GOMI. As we sat around the table trying different cookies from the Instagram-famous Utah cookie company, Crumbl, she admitted to me that she thinks Alice is pretty funny, and a good writer.

To demonstrate, Shannon recalled one of her favorite things Alice has ever written about her, on GOMI's "front page," where Alice writes blog posts about particularly juicy blogger drama. Shannon started laughing even talking about it.

"I was crying," she said as she loaded the blog post on her phone. "Alice made it even funnier than it actually was."

The blog post was about Instagram Stories Shannon had posted at the end of 2021 about Dallin cutting the trunk off of their Christmas tree with a chain saw in the middle of their living room. At that moment, Shannon thought the scene was hilarious. It was exactly the kind of slapstick, Sandler-esque humor that she finds so amusing, and she, of course, posted it to her page.

"It's Safety First at the Bird Family Christmas" is the title of Alice's write-up of the incident. Shannon could barely make it through reading the blog post out loud to me before dissolving into laughter.

"The Bird family, a gene pool that might convince you natural selection is a work of fiction, have put up their Christmas tree," she read. "Did they measure and safely trim down the tree before bringing it inside and placing it in the tree holdy thingy? Of course not. Why do that when you can simply bring a chain saw into the living room and start ripping away right there on the rug? . . . Having successfully waved their middle fingers at Darwin yet again the entire brood went out to take a picture with whatever elderly stranger the mall hired this year. Happy holidays to all! Except you, evolution." A lesser woman might be horrified and embarrassed to be mocked in such a manner, but

Shannon insisted to me that she doesn't care at all. Strangers on the internet making fun of her doesn't faze her.

Shannon has a very different approach to handling the scrutiny on GOMI from that of most other influencers. While her peers tend to ignore the forums, most never acknowledging they exist, Shannon almost seems to delight in baiting them.

Shannon, Dallin says, has no interest in portraying some idealized version of perfection. "In her head, she's like entertainment, to some extent." This has kind of warped his wife's online persona into something she is really not, he thinks.

Dallin has been approached many times by women who tell him they follow his wife's blog, and they are shocked and pleasantly surprised by how normal he and the kids are in real life. He even catches them staring before they say hello. "You can kind of tell people are waiting or watching to see something sometimes," he said.

"Is Shannon playing a character on her Instagram, then?" I asked. Shannon thought for a minute and said no because everything she posts is really happening. She's just omitting the more mundane aspects and highlighting the funny things that happen.

Dallin disagreed. "The answer is yes," he said. The version of Shannon we see online is a part of his wife, sure, but it isn't *her*.

During the day I spent with them, Dallin and Shannon had the same debate multiple times. Is Shannon egging on her online critics or not? Is she playing a character of a messy, goofy mom to troll the haters, or not?

From Shannon's viewpoint, while her online critics don't

bother her, she isn't posting for them. She has another audience: women who enjoy her unfiltered take on motherhood.

Dallin, though, sees it differently.

"She knows her audience is actually the critics," he said. "She's a quasi-contrarian in that she feeds off the hate . . . she feeds off a show."

IN MY DECADE or so of reading GOMI, I have noticed a few things influencers do that really piss off the commentariat. One is too many sponsored posts, or sponsored posts that the readers think are not for products the influencer would actually buy. Another is obfuscating part of their life, whether it's how much familial support they get, how much money they have, or even keeping personal things private. A third is not recognizing how good they have it, and not seeming to care enough about their followers.

A lot of these main critiques come back to one concept: authenticity. For influencers, authenticity is everything. Followers expect perfectly authentic experiences from their influencers, which gives them the power to recommend experiences or products with authority. It's an uncomfortable reality. Influencers are only as successful as their followers allow them to be. The trust of their audience is everything.

Sarah McRae, of the University of Alberta in Canada wrote about this dilemma in a 2017 paper that examines how travel bloggers were critiqued on GOMI, and how commenters "identify and deconstruct lifestyle bloggers' efforts to perform an 'authentic' persona."

McRae noted that performing "authenticity" is one of the trickiest things for influencers to navigate. She described "authenticity work" as "curating a persona that is aspirational, but *ordinary*, attracting followers with the narrative that the extraordinary lifestyle being presented can be achieved by the average person," noting that the most recurrent criticism of bloggers and influencers tends to stem from issues of authenticity, even if most of the commenters are not cognizant that this is what they are critiquing.

So influencers are supposed to be better than you but not too much better, the "best friend with great ideas" in your pocket, whom you could be like if only you tried hard enough. This balance is rather tricky to achieve. For example, a blogger like Rachel Parcell often gets criticized online for being too perfect, for having too much money (even if she made that money herself), and for not showing a "realistic" view of what it's like to be a parent.

On the flip side, Shannon shows her kids looking messy or getting into trouble, and her own parenting fails, and she is threatened with calls to CPS. People call her a "shit mom." So really, these women can't win.

Thus, in many ways being an influencer is like walking a tightrope. You're high up in the sky, executing an amazing feat, but one misstep could cause you to tumble down.

Authenticity is something that Caitlin has thought about a lot in recent years when it comes to her business. Mainly, how much of herself do her readers and followers deserve to know? How much does she have to share? And why does being an influencer mean she has to open up every aspect of herself to criticism from anyone on the internet?

Unlike Shannon, Caitlin's default response to the trolls isn't to troll them back even harder. It's to close herself off. When she first launched her blog, it was easier to maintain an air of mystery, and therefore privacy, to ensure the trolls couldn't hit her where it hurt. Back in those days, aspirational content was everything, and it was a good thing for a blogger to look and seem perfect. Caitlin spent hours making sure her photos looked exactly right. She credits a lot of her early success to her ability to know what made a good photo and an aspirational shot.

But Caitlin didn't have just a good eye. She had a knack for perfectionism. She would obsess over her images, crafting each blog post like the page of a magazine. In the early days, her mom took all the photos for her blog, and Caitlin had complete artistic control (she still has control, but now her brother takes her blog photos). Back in the day, among what she now calls the "OG bloggers," that was the goal. They were supposed to be perfect, professional, and aspirational. You don't read *Vogue* for photos of real life, you read it for inspiration. That's what early bloggers wanted to be too.

But soon, readers began wanting more. They didn't want Caitlin to only share with them perfectly curated images of her wardrobe. They wanted to know about her life, to hear and see everything she was up to, and they wanted her to be real, to be authentic. New bloggers began to pop up who eschewed the perfect content Caitlin is so good at. Instead, they were posting photos of themselves with baby spit-up on their shirts, or in the middle of panic attacks. Moms showed the hard parts, not just the good parts, of parenthood. Fitness bloggers posted photos of their stomachs with and without good posture, just to show how easy it is to make it seem like you have a six-pack. Beauty bloggers appeared on Instagram Stories without makeup, proudly showing off their acne scars. Suddenly bloggers weren't supposed to be just aspirational, they needed to be real.

Caitlin's followers began to tell her they wanted to see her without makeup and to hear her deepest, darkest thoughts and feelings. *Tell us more about you*, they would DM her, and *Show us more of your life*. For Caitlin, though, perfection has been a hard habit to break. After ten years in the industry, she feels like she has to reinvent herself.

"I don't think bloggers will survive if you don't show more of yourself in the industry how it is today," she said.

It's not just a habit, though. Caitlin has been burned by sharing herself in the past and is now trying to tear down walls of self-protection that she built long ago to protect herself from the internet. In many ways, she was unprepared for the toll her career would take on her.

In July 2014, Caitlin made her first appearance on GOMI. Her forum now has more than 450 pages (Shannon's is nearing eight hundred), and at first, commenters mostly debated if Caitlin was really as sweet as she seemed or was a "secret mean girl." They also debated—what else?—her authenticity, because she had started making money from her blog.

"This is why I don't read her blog anymore," wrote one. "Trust her reviews or pretty much anything that she says. That girl is fake and blogs only for the money."

Then, commenters began to take aim at Caitlin's relationship with Chris. They focused on a photo shoot she had done with him around Christmas and posted on her blog, gushing about the local Charlotte photographer they used. The photo shoot is achingly 2014, with Chris in a red flannel scarf and Caitlin in a red slouchy beanie. On her feet, as she put it, were "my red Hunters—y'all know I love those."

For the commenters on GOMI, this photo shoot was not it, and they began to question other aspects of Caitlin and Chris's relationship. They derided her as desperate for having a photo shoot with her boyfriend before they were even engaged and accused her of misleading people about how long she had been with Chris, implying that she was acting like the relationship was more serious than it was.

One wrote, "i find it really inauthentic of her to write on IG that she's celebrating her two-year anniversary with her bf. she blogged earlier this year about how she and her bf were broken up for a few months . . ."

The commenter's examination of Chris and Caitlin's relationship then began to expand. Using the "Life Updates" post, and another one in which she alluded to the breakup ("I also went through major heartbreak . . . but because of it, I learned just how strong of a person I am," she wrote), they began to speculate about what exactly was going on behind closed doors. They sneered that Caitlin was desperate for Chris to propose, and speculated that the relationship was one-sided. They even joked that, like Kate Middleton, she was a "Waity Caity."

"Probably a bad sign when your boyfriend doesn't want you to be around when he needs people around him," wrote one. "Sounds like she is super stressful and pushing for that ring. During their breakup, she moved from Greenville to Charlotte most likely to be closer to him and to try to get him back. I feel bad for this guy, and he needs to run fast."

One day, Caitlin stumbled upon what people were saying about her online. They wrote that they hated her hair and her eyelashes and speculated she had an eating disorder. Worse though, she felt, was they wrote about her and Chris. Some people speculated about why they had broken up, saying that Chris probably cheated on her, and other things that Caitlin knew weren't true.

Unlike Shannon, Caitlin cared about what these anonymous people online were saying about her. She cared a lot. For months she obsessed, reading all these things strangers were saying about her. Sometimes, she would just read them and cry.

She was afraid to post anything, concerned suddenly about what all these anonymous people were saying about her. She found herself posting things to try and appease these faceless online commentators, and realized she could never survive in the industry if she continued to play to the whims of the internet.

Caitlin stopped reading the forums cold turkey and now refuses to look at them. That doesn't mean she doesn't get hate—people send her nasty DMs all the time—but she refuses to let snark forums impact her career. She also found a sympathetic ear in some of her blogger friends. One of them told her something that has stuck with her to this day.

"The fact that people are so interested in you that they're digging into your relationship, they're digging into your personal life, you know, it hurts," she remembered them saying. "But in a way, if you look at a football stadium, there's the majority of people cheering for you and loving you. And then there's this crowd, this one section of the stadium, where they don't like you and despise you. And my friend is like, 'Pay attention to all those people that are rooting for you and cheering you on.'"

Over time, Caitlin has developed a thicker skin. When she gets nasty messages, she tries to tell herself that in some weird way, it's a good thing. She tells herself, "If I have people this invested in me that they hate me so much and that they're digging into my life, then it must be a sign I'm doing something right to even have haters."

But Caitlin acknowledges that this early experience has made it difficult to adjust to the new era of influencing, in which everyone is expected to be "real." In 2014, when she watched the internet pick her relationship apart in real time, she had made a decision. She needed to protect herself. From then on, *Southern Curls & Pearls* would be "strictly fashion." She wouldn't talk in-depth about her relationships, her feelings, or personal issues in her life. She would not be hurt like that again.

Of course, she has loosened up over time, sharing more and more of her life. But there's a part of her that still wants to protect her heart, and is nervous about what will happen if she lets the world all the way in. Unfortunately, it is the cost of doing her business. She has never considered quitting over it, though, and insists that all the heartache has been worth it. She has grown as a person and as a business owner.

"I'm thankful that blogging brought that to me because I feel like I'm thick-skinned now in all aspects of my life," she said. "I'm still emotional when it counts that I just don't let negative stuff bother me anymore."

It's easy to regard Caitlin's approach to the internet commentators as the right one. Why shouldn't we ignore those who hate us, and shield ourselves from criticism? After all, what right do the commenters have to dissect her life, and that of her boyfriend?

But how does this impact Caitlin, and her followers?

5

It's just a fact. Being an influencer has made Caitlin a lot of money.

Well, she would describe herself and Chris as "wealthy." When she says the word, she does so in a hushed tone, as if she's telling me a secret.

Caitlin is conflicted. On the one hand, she wants people to understand how successful she is, and her income is a testament to that. After all, she built her wealth from her own business. That's not to discount her privilege of a comfortable upbringing, or Chris's full-time accounting job at Hanesbrands' corporate office in Winston-Salem (he works on the Champion line). But it's Caitlin's income that has catapulted her family from comfortable to wealthy.

Right now, she says she and Chris are saving about 90 percent of their income, and that's after living a lifestyle that seems luxurious, if not downright lavish. They have a beautiful house in Winston-Salem (with an extremely enviable closet) as well as a beach condo. They take picture-perfect vacations to Italy and Greece. They have a life that many would kill for—on, she says, 10 percent of their income.

Though Caitlin is a self-made woman who created a life for herself beyond anything she really expected, few others understand or acknowledge that fact. Many assume she's a trophy wife. She gets questions all the time about what her husband does. *He must have an amazing job—just look at your nice house and clothes and Louis Vuitton handbags* (handbags are Caitlin's big luxury splurge). The idea that Caitlin is contributing to her household income in a significant way doesn't really cross their minds. Others ask her: *What do you do all day?* "Um, this," she wants to respond. "It's my job."

But she bites her tongue, worried about the impact that full transparency, or living a "crazy lifestyle," as she puts it, would have on her career. In many ways, she almost hides her wealth from her followers. She doesn't want anyone to really know how much money she has.

"I don't want to flaunt it. And I don't want that to be, like, the perception that I'm giving off, like just this rich girl. . . . I want to stay relatable."

Caitlin's quagmire is a unique pitfall of her chosen career. She is hugely successful at what she does, but she feels if she fully indulges in the rewards of her labor, she could lose what makes her successful in the first place. Sure, some influencers do flex for the 'Gram. Caitlin has watched as her peers in the industry, who used to wear Forever 21 and drive normal cars, have changed.

"All of a sudden now they're, like, driving an Escalade with ten billion Chanel bags and a mansion," she said.

This is why she is so careful about how she presents her public image. She may be able to afford luxury goods that her audience probably can't. If that's all she posted about, they would seek out some other woman who is selling them things they can actually afford.

Authenticity for Caitlin is everything. To be completely honest, she doesn't even want a closet full of Chanel. Most of the time, she *likes* shopping on Amazon or a small boutique and finding a good deal or something really flattering to share with her followers.

"I feel like normally you wear an outfit a couple of times and then you don't wear it again. So there's no point in spending, like, a thousand dollars on a dress if you're going to wear it just one time," she said.

It's a weird situation to be in. For Christmas 2021, Chris gifted Caitlin a pretty amazing present: a Mercedes G-Wagon. She posted about it on Instagram and most of her followers were thrilled for her (if not a bit envious). But on snark forums, what Caitlin predicted came true. They immediately blasted her as out of touch.

I'm often fascinated by how insanely curious everyone is about how much influencers make. It's the number one question I get asked about influencers, by far. While writing this book, I asked a few times on my Instagram what was the biggest question about the influencer industry people would want answered. Almost every question was about money. *How much do they make a year? How much do they make per post? Do they save their money? How do they do their taxes?* And on and on.

I had been thinking a lot about these questions before talking with Caitlin about her earnings. Why did everyone want to know so badly? And would she even tell me? I felt awkward asking, but I knew I had to.

I gently nudged her to tell me, but she didn't take the bait. She did, however, weigh in on why she thinks everyone wants to know so badly.

"First, I think people are always thinking about doing it themselves and they're like, 'How much money do you make? Is this something that I should get into if you can make a lot of money?'" she said. "And then I think a lot of people have watched influencers kind of start from nothing. And then all of a sudden it seems like they have so much, and they drive a nice car. They're naturally curious, like, 'How are you able to do that? How much are you really making from a blog?'"

Caitlin's followers feel they *deserve* to know how much she makes because they have such an intimate relationship with her career. They know that the money she makes is because they trust and respect her enough to follow her account, swipe up on her links, and use her promo codes. *We made you*, they seem to be saying. *You owe us transparency.*

So, I set out to try and provide some transparency on this

industry that people find so opaque and confusing. The TL;DR: There's money in influencing. Like, more than you would expect. Like, take what you think your favorite influencer may be making and double it. Then, maybe double it again.

After all, experts predict that in 2022, influencer marketing, or the economy around brand partnerships, will be a $16.4 billion industry. Most influencers have one or more other revenue streams as well. Caitlin, like many influencers, also makes commissions from LTK, Amber Venz Box's company, which helps them make commissions when followers click on their links. In 2021, Caitlin added another revenue stream by creating her own line with the clothing brand Pink Lily.

It's not just the influencers who are making bank, though. The influencer industry has created an entire ecosystem, creating jobs for thousands of people across the country. Influencers have assistants, managers, and agents. Some have consultants who help them create content. Some have an employee who literally just answers their DMs.

Much of this industry sprung up organically, largely driven and created by the influencers themselves. For many years, there was no way for influencers to earn money directly through Instagram, which has had a distinct effect on the way the entire influencer economy has developed.

Compare Instagram to YouTube. From the beginning, YouTube has actively cultivated its creator community, allowing them to make money directly off of the platform for years via ads and more, based on how many views their videos got. When the creator economy was in its infancy, YouTube developed a "partner program," in which the company cultivated its creators via support and revenue sharing. In 2011, YouTube trumpeted in a blog post that many of its fifteen thousand "partners" were "making enough money doing what they love to buy a new house or even make a career out of their videos," including "hundreds" making six figures in YouTube revenue a year. In 2011, YouTube invested even more in its partnerships, offering credits to upgrade camera equipment, meetups, and workshops, as well as revenue and support.

In the years since, YouTube has only invested more in actively cultivating its creator economy, truly treating the people who create content for the platform as partners in its business. From big things, like putting huge YouTube stars like Hannah Hart and Lilly Singh in ad campaigns and on billboards, to smaller gestures, like sending plaques to creators who reach one million subscribers, it feels like the platform supports its stars. To create on YouTube is not only to be able to earn money without seeking out a third-party partner (although, of course, many also do), but to truly feel like you are working in tandem with the platform, rather than against it. There's an argument to be made that YouTube's advertisement of its talent in the media and on billboards across the country also has led to YouTubers, in general, becoming much more successful in Hollywood and more famous in popular culture than your average Instagram influencer.

Instagram couldn't be more different. For many years, as influencers built the platform into a lifestyle, fitness, and shopping hub, Instagram had a very laissez-faire attitude toward them, but happily got rich off their content. It is the content that drives people to open the Instagram app every day and scroll, but it is Instagram that reaps the benefits of those eyeballs. The company runs ads on its platform that it takes all the profits from, which has led to a ton of wealth generation for its parent company, Meta, and its shareholders (Bloomberg reported in 2018 that Instagram had crossed the $100 billion mark in revenue that year).

Despite the fact that creators drove so much attention to Instagram, the platform has been reluctant to share its spoils. Until 2021, there was no way to make money directly through Instagram (enter entrepreneurs like Venz Box), and influencers had no guarantee or assistance in ensuring that their paid advertisements, which they negotiated separately from the platform, would be seen or remain in place once they uploaded to the platform. Many influencers have told me over the years that they don't receive any special support or troubleshooting from Instagram if something goes wrong with their account; the level of help they get is similar to what any average person with an Instagram account would get. Literally, they say, the level of support is the same as that for the personal account with two hundred followers you use to post photos of your dog.

For creators like LTK founder Amber Venz Box, this disconnect has never made much sense.

"If [we] don't put the content in your pipes, you only have pipes," she said of herself and other influencers. "They don't value the creator in the ecosystem."

In the summer of 2021, with great fanfare, Instagram did slowly begin to announce ways to make direct revenue on its platform, introducing bonus programs and initiatives that would allow influencers to make money through sponsorships arranged on the platform directly. They committed $1 billion by the end of 2022 to these programs. But many of the first bonus programs were geared toward Reels, Instagram's TikTok clone. Traditional Instagram influencers didn't feel that Reels fit their content or their audience. The move, to them, felt like Instagram was merely trying to capture a new audience of young, Gen Z TikTok stars rather than reward the influencers who had been working on their platform for years.

Thus, in the absence of Instagram's direct support for influencers, the industry evolved on its own. Influencers were largely left to their own devices to figure out how to implement brand partnerships, how much to charge, and how to ensure they stayed within Federal Trade Commission guidelines on ad disclosure. Companies like management agencies for influencers and platforms like LTK sprung up to fill the void. Caitlin's manager, Kirstin Enlow, has made it her career goal to make sure her clients, like Caitlin, are paid what they are worth.

"While for some people, it's baffling that influencers would get this much money, it's also baffling to me that influencers aren't getting as much money as some celebrity endorsements because I know what those types of deals look like," she said. "And I'm like, 'Our creators should be getting paid that much.'"

BEFORE BECOMING AN INFLUENCER, Mirna had never mountain biked. Now she does it all the time.

Turns out, she loves it. But learning to mountain bike wasn't her choice. In fact, she had done some biking in the past, trying out a beginner bike course that, in hindsight, had a lot of nerve labeling itself for beginners. She kept falling, and she hated it. She vowed to never again try biking as a sport.

Then, a sponsor came calling. They wanted to work with Mirna, to pay her to post about their bike.

"They're like, 'We're sending your bike to you. Where are you going to ride?' I'm like, 'Well, first I need to learn how to ride it,'" she said with a laugh.

Mirna had a friend who is a mountain bike instructor in Arizona, so she decided to head down there to do what she called a "mountain bike intensive." It worked out because her brother lives there and she was able to bring her mom along. They got to fly first class and stay in a nice hotel. And the trip was for work, so it was tax-deductible.

"It was, like, super, super positive, except for when I got impaled by the seat," she said with a hearty laugh. "But it was really cool."

Since then, Mirna has fallen in love with mountain biking. Even though she still frequently falls, she has found the sport to be an apt metaphor for life.

"Mountain biking continues to be scary to me, but the more I fall, the less fear I have," she wrote in one Instagram post, alongside photos of her beaming as she bikes through the Vermont foliage. "I like to think of this moment as a representation of how I'd like my 46th year to be."

Mirna has fallen in love with other sports through sponsorships too, like skiing. When she was first approached to do sponsored content in which she would need to ski, she told them that she wanted to do it, but only if she could actually learn how to ski. Instead of getting paid for the partnership, she negotiated in her contract a ski pass for the season and a season's worth of lessons. Now she is an epic skier, and skiing is one of her favorite things about living in Vermont.

These partnerships are perfect for Mirna's brand. Her followers get to watch her do things she genuinely enjoys—being outside and trying new sports—and thus they don't feel like ads. Because the content is genuine, followers like and embrace her posts about skiing and biking, even if she includes #ad at the end. It's a perfect balance and has even helped her create genuine relationships with her followers.

"There are at least two people that sent me stories where they were like, 'You got me to get back on the slopes,' or 'I took a ski lesson this season and I, and I really loved it,' or 'I think I'm going to go out on a bike.' So that's been really cool," she said.

This marriage of genuine content and brand partnerships is the way the industry is supposed to work. Influencers have a brand, and they seek out companies that create products that align with their brand. The influencer gets to try something new and make money, and the company gets to sell to the influencer's audience, who are likely to buy the product because they trust the influencer and have similar interests. These marriages between companies and influencers via brand partnerships are one major revenue stream for influencers and are making a lot of people very wealthy.

In early 2022, the market research company Influencer Marketing Hub predicted that the industry would grow to 16.4 billion that year. Although new platforms for sponcon, like TikTok, are on the rise, as of 2022 the majority of companies chose to market on Instagram, followed by TikTok and then YouTube. On average, brands are spending around $174 for every paid partnership you see on your Instagram feed. That's not a lot of money, especially compared with what a celebrity charges for an endorsement.

For companies, these campaigns are incredibly lucrative. According to Neal Schaffer, a marketing researcher who specializes in social media, studies have found that influencer marketing gives companies eleven times the return on investment that banner ads do, and businesses are earning a little more than five dollars for every dollar they spend. This is likely because of the relationship influencers have with their audiences. According to Schaffer, when surveyed, 75 percent of consumers said that they trust the recommendations of social media influencers, as well as their own social circle.

In fact, 40 percent of millennials said they trusted the recommendations of influencers over those of their own friends and family. Trust in influencers is likely to grow, as younger people are more likely to trust influencers than older generations. For example, 70 percent of teenagers surveyed said they trust the recommendations of YouTubers more than celebrities. According to Influencer Marketing Hub, "a majority of consumers prefer following influencers who look and act like us, instead of celebrities."

A lot of the discourse around the influencer economy tends to focus on how much the influencers themselves are raking in. But it's important to note that another big winner here is actually the companies themselves.

Influencer marketing is a great deal for brands. Celebrity endorsements are expensive. Kylie Jenner, for example, is rumored to have made $1 million in 2016 as a spokesperson for Puma, Brad Pitt has reportedly earned $7 million for hawking Chanel, and Justin Timberlake $6 million for advertising McDonald's. Lisa Rinna reportedly earned $2 million for doing a commercial for Depend undergarments!

There are of course differences between a celebrity endorsement deal and influencer sponcon, but the latter costs companies much, much less, for an extremely high return on investment overall. Influencer sponcon is also much cheaper to make. The costs of filming a traditional commercial can be staggering, but companies can pass production costs on to influencers. That means they don't have to pay for a set, costumes, makeup artists, a film crew, or anything but the finished product. They pay the influencer, and the influencer handles the entire production. The brand weighs in on the style or theme of the ad and sets specific things they want, but it's up to the influencer to produce it. And yes, even a simple Instagram photo can be a production.

Many influencers hire a photographer to shoot at least some of their photos, or they have assistants who help out on shoots. Caitlin's brother, for example, takes all of her Instagram photos as a large portion of his full-time job as a photographer (he also does other photography jobs, including weddings). But few, if any, Instagram influencers are working with a full production team comparable to that of a commercial shoot, and many shoot their campaigns on their own. Mirna sometimes enlists her teenage son to help her shoot campaigns, and occasionally hires professional videographers and photographers. But she shoots many of her campaigns herself.

I got to accompany her on a shoot in early September in Vermont. It was still pretty hot for the fall, and the foliage in the rolling hills was just starting to change colors. Mirna had struck a deal with the Nature Conservancy to promote its programs on her Instagram feed and Stories. As part of her sponsorship, the conservancy had booked two days of showing Mirna some of their offerings. Today was a bird-watching tour, hosted by Jim, the director of conservation initiatives. The group's head of communications, Eve, and Heather, its director, tagged along.

I had come prepared in hiking boots and leggings for our excursion, which I was fascinated to observe. After all these years of reporting on internet culture, I had never actually watched an influencer create content for a campaign in real time. It was kind of funny to me that my first time watching an influencer in action was this specific campaign. When most people picture Instagram influencers making money, they imagine thirst traps and Sugarbear Hair, not the Vermont woods and learning about birds.

Eve, the communications director, excitedly told me that she had been pushing the Nature Conservancy for a while to try new types of marketing, like working with influencers. Mirna's was the first influencer campaign they had approved. When she had been leaving the house that morning, she had told her son she was meeting with a real-life influencer. "He didn't believe me," she said with a laugh, so she whipped out her phone and showed him Mirna's profile.

The hike was a delight, and while we didn't see as many birds as we had hoped due to the season, I felt invigorated by the nature and by Mirna. People often ask me why people follow influencers, why they buy things from them, and why they care about their lives. It takes a certain amount of charisma to be an influencer, and it's a skill that not everyone possesses. But Mirna makes it look easy.

She asked deep, probing questions and exuded genuine excitement about bird-watching and the conservation of Vermont. I know that she genuinely loves Vermont and new experiences, but she was on the job—this wasn't just a nature walk with pauses for photo ops. She took photos and videos almost constantly, often stopping us to make sure she got the shot and directing Jim to tell her followers some fact or other about the birds. Mirna was so engaged, it raised the excitement and passion level of the entire group. Jim clearly relished having such an eager pupil in Mirna and gave us a stellar tour. When he told us to whip out our binoculars to try and spot a bird, we did so with the enthusiasm of superfans at a concert.

At one point, I pulled Mirna aside. I asked her how she did it. Did she get tired of meeting with people like this day in and day out, and did she ever have to fake it?

Mirna said no. She was in her element, learning from others in the outdoors.

"I like seeing people be engaged and reading off their energy," she told me.

In Mirna, the Nature Conservancy was getting a great deal. She's selling their brand to more than a hundred thousand people, and she's doing it for a fraction of the cost of hiring a film crew to shoot a commercial. On top of that, her audience trusts her, and they have interests similar to hers. If they are local, they may go check out the trail we hiked (one of the goals of the team was to get more visitors to the trail, which had recently been renovated). If they aren't, maybe they will check it out next time they are in the state.

Well, I was sold on the idea that bird-watching in Vermont was the most fun and interesting thing ever. Simply put, Mirna is a great salesperson. I decided to book a trip to Vermont for my mother (an avid hiker and bird-watcher) and me for Christmas, and even looked into doing one of the Nature Conservancy's tours. Sure, I wasn't influenced in the traditional sense, as I experienced it IRL, but I never would have done it if not for Mirna.

It took Mirna awhile to figure out how to run her influencer business, but at this point, she is pretty comfortable with negotiating her own contracts with companies who want to work with her. She's amazed, as many influencers are, at how unregulated the industry is and how widely disparate rates and payment can be, and she said she has had to learn to make sure she is not undervaluing herself when negotiating with companies.

"What is my value to them?" she said she now considers. "Why are they asking me this? Clearly, they see some value in asking me to do this, because maybe I have a growing following, maybe I have really awesome engagement, et cetera. So those are all the questions that you have to ask yourself."

The more established an influencer becomes, the more willing they are to stand up and say no to companies that they believe aren't compensating them fairly. When Mirna gets a request for a partnership, she takes into account how big the company is, and the value she believes she can bring it based on her unique story, her follower base, her engagement rate, and her proven ROI from other campaigns. Some of the biggest companies, which she knows have enormous annual revenue streams, can be the stingiest. In negotiating the deals, Mirna now leverages her engagement, her dedicated following, and her experience as a motivational speaker. Her base rate is currently around $2,000 for a post and $2,500 for a series of stories, but she has charged up to $10,000 for a campaign involving multiple elements.

The confidence to set what you believe are fair rates comes with time, though, and many smaller influencers, especially creators of color, can get paid far less than what they are worth. For many years, it was standard practice in the industry for companies to sometimes only offer free products in exchange for campaigns, which cost the companies essentially nothing. Mirna and many other content creators regard these offers as insulting.

"You need to charge them something so that they know you're serious, and so they won't keep taking advantage of other people because you're actually working," she said. "When you go out and you take a photo and then you write a whole caption, you're spending time on that. You're doing work for them."

As the industry has matured, many influencers like Caitlin have also hired professionals to help them make sure they are getting a fair shake from the brands they work with. Enter talent managers.

Many big influencers have talent managers just like Hollywood celebrities. In fact, the agency that represents Caitlin, Digital Brand Architects, is owned by United Talent Agency, one of the biggest agencies in Hollywood. It's a new industry, but a growing one. These managers are just one industry that has popped up in the influencer economy, adding to its growth. As Influencer Marketing Hub reported in 2022, the influencer support industry is growing rapidly, and the number of influencer marketing–

related companies and services increased by 26 percent in 2021 to become more than 18,900 worldwide.

Kirstin never imagined she would one day be negotiating brand deals for influencers like Caitlin. When she went to college, she had dreams of working in the music industry as an agent. She wound up working in the industry in a different capacity. But then in 2015, she ran into an old boss she had known as an intern. He had pivoted to representing YouTube talent and told her she should consider joining him. Then he told her just how much money his clients were making.

"I was just flabbergasted because I was coming from the music industry where just to make money on a song was like pulling teeth to get anything from it," she said. "I was like, 'Oh my God, this is . . . where the opportunity is.' And so I've completely pivoted."

In the years since, Kirstin has moved up to become a director at DBA, where she manages a stable of content creators she describes as a "mixed bag." She represents fashion and lifestyle influencers like Caitlin, as well as many plus-size creators, like *Sports Illustrated* model Hunter McGrady and blogger Gabi Fresh. She's passionate about representation and fighting to ensure all her clients—such as the hijabi influencers she represents—are fairly compensated.

One of Kirstin's goals is to make sure her clients like Caitlin get paid what they are worth. Figuring out what that is, exactly, is complex. Kirstin and her colleagues examine every potential deal separately and set the rate based on a variety of factors. She attempted to break it down for me as best she could.

Say you're a creator with five hundred thousand followers. That follower count alone would usually equal a five-thousand-dollar rate, minimum (and they hardly ever make a deal that pays that little, she said). Then, she adds in all the other factors. How's their engagement rate (an algorithm that takes into account how many followers interact with the influencers' content) on average? If it's good, like 3 or 4 percent, that adds to the minimum. If it's excellent, like up to 20 percent, that adds a lot more. Kirstin sees a lot of Gen Z creators whose engagement rates are off the charts, and that matters. "Some of these Gen Z creators' engagement rate of that scale would be equivalent to someone who maybe has a hundred thousand or a million followers," she explained.

Next, how popular is the influencer? If they are getting so many brand deal requests that they can afford to turn down many if not most, as Kirstin said Caitlin does, they can be more selective. They can raise that rate even higher because they don't go with just any brand. The time and resources the influencers need to commit also matter. How many hours do they have to work on the content that goes into this deal? Are they hiring a photographer and if so, how much are they paying them? That gets factored in too.

All in all, Kirstin said a successful influencer with a healthy follower count and a good engagement rate should be able to easily charge ten thousand dollars per brand deal. She thinks anyone who thinks that's absurd doesn't understand the business. Compared to a celebrity endorsement, that's chump change, especially considering how good a deal companies are getting.

"If you think about these budgets for a lot of these brands, sometimes the brands are working with million-dollar budgets and in comparison, it's actually pennies to what they're paying to have a commercial if you think about it," she said. "And the eyeballs that they could maybe get from a commercial are potentially the exact same amount of eyeballs that they get from a creator."

While the industry has matured from the early days, Kirstin still thinks there are a lot of misconceptions that lead people to undervalue them. "I think that there's a lot of taboo that exists with the influencer space, unfortunately, and people don't want to look at influencers as a positive thing," she said. "There's a lot of negative connotations with influencers, like they're adding negativity to the world . . . versus they're actually doing a lot of really good."

She also pointed out that it's not like anyone can just start posting to Instagram and making ten thousand dollars per post. To reach that echelon, you have to have been working for years, and more likely than not you were way underpaid for years before finally being able to charge what your content is actually worth.

"Maybe the first three or four or five years of your career, you weren't making any money and you were grinding every single day all day long to get to that level . . . so it's also kind of making up for all the lost time," Kirstin said.

Influencers don't get to do just any ad they want, though. They must ensure that the product they are selling aligns with their brand, something that sounds simple but is actually incredibly difficult to navigate.

No one really thinks that Jennifer Aniston uses Aveeno, and frankly, no one really cares. But if Caitlin or Mirna does an ad for a product that her followers think even slightly doesn't align with the image she has cultivated, it can be disastrous. While many actors sell products just as much as influencers do, they are seen as having a "real" job or talent. The entire job of an influencer, however, is to influence us, and their recommendations are only as good as the trust their followers have in them.

Caitlin has written extensively about her desire for the products she uses to be "clean," free of harsh chemicals, and as organic as possible. So now anything that she sells needs to be "clean" enough, or her followers instantly home in on it and accuse her of inauthenticity. Once, she did an ad for a perfume and her followers grew upset, saying they didn't believe Caitlin actually used the product because it had ingredients she likely wouldn't use. The thing was, they were right. Caitlin realized she hadn't really used the product much since taking the partnership, because she didn't love the ingredients. When the company came around the next year to ask if they could work with her again, presumably because they were happy with the first campaign, Caitlin declined. Her followers wouldn't like it, and she realized she wouldn't be authentic if she took the partnership.

"I shouldn't have done it in the first place, because I want them to be able to trust me when I give them a recommendation that they're going to love it as well, because I'm putting my stamp of approval on it," she said. "And I just think it can just ruin your brand if you promote anything and everything."

Therein lies an interesting feature of the unrelenting feedback that influencers get from their followers and others online. Many times, it is valid. One of the earliest complaints about influencers and bloggers on GOMI was that they were getting too greedy, and were taking campaigns for products that they clearly had never and would never use. Many of these commenters were right. Influencers *were* doing that, and it damaged their credibility and the credibility of the entire industry. Snarkers on sites like GOMI were some of the first checks and balances on influencers, in an industry where there was little regulation and no one seemed to be paying attention to it.

Caitlin has seen her peers in the industry fall into the trap of just taking every offer that comes their way, and become cautionary tales.

"I've seen it go badly because then no one wants to order whatever you're promoting because they're like, 'Well, I ordered this other thing and I hate it. She's just doing that for the money.' It just gives a bad taste in everyone's mouth," she said.

For most influencers, though, this isn't just about appeasing the peanut gallery. It is more personal than that. Their brand isn't some abstract thing they created, it is a reflection of their actual beliefs and personality. Caitlin genuinely cares about clean products in both her personal and professional lives and appreciates her followers holding her accountable.

Mirna is also extremely careful to accept only partnerships that align with her brand. She is passionate about the causes her brand champions, and she is dedicated to aligning with partners who want to help her fulfill this goal.

One partner that does not fall into that category? The dieting company Jenny Craig, which emailed Mirna as we sat eating lunch after our bird-watching trip. Mirna saw the email and was disgusted. She was horrified that anyone from Jenny Craig would ever reach out to her.

She read me the email. They were looking for someone to partner with who could help people adopt so-called healthier lifestyles after being cooped up for months during the pandemic. Mirna rolled her eyes. They must, she said, not actually have known anything about her or her platform. Otherwise, they never would have sent her something so insulting. Mirna is happy in her body and doesn't want to change it. She doesn't go on crash diets or exercise to lose weight. She is trying to teach her followers that you can have a body like hers and be an athlete. She would never accept an endorsement from a weight-loss company like Jenny Craig, no matter how inclusive they are trying to be.

She stewed on it for a second as we ate our salads. Usually,

Mirna would have her assistant, Kimberly, or Margaux respond to an email like that and decline, but she planned to respond to this one herself, to give them a piece of her mind.

Even the biggest companies must prove to her that they are genuine in aligning with her values and goals. She doesn't want to be used as some sort of token for companies to prove they care about plus-size or Black people when their actions say otherwise.

Take Lululemon, for example. When the company reached out and expressed interest in partnering with Mirna, of course she was flattered. After all, Lululemon is one of the biggest women's sports apparel brands and, she noted, is considered pretty "luxury" as well.

But when she got the initial inquiries, she told Margaux the answer was going to be no. She hadn't liked what she had seen from Lululemon in the past. In 2013, the company's founder and then-chairman, Chip Wilson, infamously blamed complaints about quality control issues on the fact that he felt bigger women weren't built for the clothing.

Wilson resigned as chairman after the comments drew a heavy backlash in the press, but Lululemon's reputation as a company that didn't care about anyone bigger than a size 6 remained for several years. When Mirna heard Lululemon wanted to work with her, she immediately thought of Wilson's comments.

"They had a really not-so-great, or rather awful reputation with plus-sized people," she said.

Mirna asked Margaux to handle the correspondence. However, as her publicist chatted back and forth with the company, Mirna was surprised. Rather than accepting her no and moving on, Lululemon's representatives engaged with Mirna's concerns and answered several of her questions in a way that made her feel they actually may be doing the work.

So, she finally agreed to a Zoom call with the representatives. She laughed when she described the meeting.

"My game face was on. . . . I have my notepad. And I was all business," she said. Mirna prepared to ask the representatives hard questions about what they were actually doing to make their brand more inclusive, and to determine if she could actually represent them to her audience with a clear conscience. The meeting went better than she had expected.

"I finally smiled at the end," she said.

Mirna was impressed. Not because Lululemon was perfect. They didn't have answers to every one of her questions. But she felt that they were trying. When she asked them hard questions about representation, they listened and seemed like they actually cared what she was saying.

Still, she wasn't totally sold.

"My concerns were, what's in it for me? What's in it for my community? What is my community going to think of me if I partner with Lululemon?" she said. She didn't want anyone to think she was letting her principles slide just because a fancy brand came knocking on her door.

What finally convinced her was the clothes. When Mirna got some samples in her size, she had to admit she was impressed.

"I wore them and was like, 'Oh, actually they have put some

thought into this plus-size sizing,' because, you know, not everybody does," she said.

It was only after all that, and plenty of negotiations and calls, that Mirna felt comfortable signing on with Lululemon. She is proud of the deal, not just because it gave her a new platform and visibility with a major apparel company, but because she did it right.

Since signing with the brand in 2021, Mirna has met with the entire executive leadership team to talk about their plus-size offerings. For example, she has told them that they don't have enough options for bigger athletes. Much of the clothing they offer in her size doesn't come in as many fun colors as some other sizes do. Mirna told them she doesn't want to just be in black all the time, and neither do other athletes her size.

Being able to be an advocate for her community is one of Mirna's favorite parts of the job. As she was negotiating with Lululemon, she decided to go directly to the source to see what concerns her community had. She wrote in her Facebook group that she had "the ear of one of the largest female apparel companies," and asked her group members what they would say if they were in the room. What did they need? What made them feel confident? And how were they not being served by athletic apparel brands?

"That thread was very long; it was very indicative of how much pain people in the plus-sized community have whenever they're trying to buy a simple piece of fucking clothing," she said.

Mirna then took those complaints and desires straight to Lululemon, who, she said, seemed genuinely interested in fixing the problems. Of course, it is a win-win for them. Not only is Lululemon getting a great deal on advertising by partnering with Mirna, but she's also showing them how to reach new customers and even running focus groups with her followers for them.

"The fact that I get to say that to them is really cool," she said. "And really, that sends a message to me and to my community that, like, people are actually listening."

It is in this way that Mirna feels she is truly making a difference, and it feels right.

"You have to build trust. We have to build community. You have to make your customers feel like you are listening to their needs, that you are aware of their needs and their desires," she said.

In mid-2022 as I was wrapping up reporting for this book, I got the latest issue of *Runner's World* in the mail. The first two pages were a two-page spread for Lululemon. The model? Mirna Valerio, the new face of the brand.

6

On January 28, 2020, Shannon and her kids were home alone. Dallin was on a work trip. Her youngest child, London, was only six weeks old and her son Brooklyn had recently broken his leg. That night, she found herself unable to produce any breast milk. She was taking medication that made her supply decline. When she realized she had no formula or saved breast milk, she grew desperate to feed her hungry baby but didn't want to rouse all her kids to bring them to the store with her. After calling some friends and neighbors, around two a.m. she called 911.

She knew the officer who responded. When we were wandering the neighborhood on my "Mormon blogger tour," we even saw him driving by. According to Shannon, the police officer usually posts up on their street and stays there all night waiting for a call. It's kind of funny, she says, because there's not much going on in Alpine, so he spends many nights just chilling. Shannon often chats with him when she goes to get her mail.

So, when she began to rack her brain for who may be up at two in the morning, she immediately thought of the police officer.

"It didn't even faze me in a way," she said. "I was like, 'I know who's awake!'"

The officer came through for her, buying baby formula and delivering it to her house in the middle of the night. Shannon was grateful to him and decided to share the saga on her Instagram Story.

"I legit tried calling/texting people and no one was answering at 2am and I couldn't leave my kids alone or drag all of [them] to the grocery store at 2am . . . I am so grateful they (the police officer) helped me . . . they were my angels. I was so miserably scared I had legit never been in such a crisis," she wrote in part on Instagram.

In Shannon's mind, the story was both her typical, Sandler-esque goofy fare (silly her, ending up in this situation) and also a feel-good story about a nice cop doing good in her community.

She never expected the story to go viral. However, it was catnip for local news stations like KSL in Utah. (Local news loves a "good cop doing a good deed" story, which is controversial, to say the least.)

"It was a 911 call that officers had never received before: a Utah mother in desperate need of baby formula for her newborn in the middle of the night," wrote KSL in their story on the case, continuing, "Lone Peak Police Department's public information officer applauded his fellow officers' actions that night. . . . While it was unique, it wasn't much different than an officer helping change a flat tire. . . . Bird couldn't thank her middle-of-the-night helpers enough and was glad the responding officers were parents, too, and understood her plight."

The story then spread like wildfire. Shannon was featured on CNN ("As a mother of five young children, Shannon Bird said she considers herself somewhat of a pro at the baby-raising game," the story reads) and outlets as far away as Chicago and the UK.

At first, the attention from the media was kind of cool. Shannon described it as a "whirlwind," ticking off all the shows that contacted her and excitedly telling me she got a free trip to New York to do interviews. She had producers "pounding on [her] door" asking her for exclusives. For a minute, everyone seemed to want to talk to her.

Then came the backlash. Online, Shannon was painted as the epitome of a clueless white woman, using her privilege to call upon law enforcement as her personal errand boy. Many questioned how a mother of color would have been treated by police in this situation (probably very differently). People called Shannon a neglectful mother, pathetic, and an attention seeker, and accused her of perpetrating a publicity stunt.

In retrospect, Shannon says she didn't really think about the implications of what she was posting. In her mind, she wasn't taking resources away from her larger community. She figured her local cops likely were not out responding to a crime in the middle of the night.

"I was like, 'Wait, you're the ones bringing race into this, I didn't think it was a racist thing at all.' That's just because I really am color-blind. . . . I didn't know my white privilege, I guess," she said.

This decision to post about the cop and the formula has had a profound impact on every aspect of Shannon's life since and has radically changed her perspective on both her life and her career as an influencer. It's constantly on her mind. Even two years later, in January 2022 when I visited her, she brought up her 911 call within the first ten minutes of my arrival and referred to it constantly afterward.

The most serious and devastating impact it had on the Bird family was the real-world one. Shortly after the incident went viral, Shannon started getting more hate than she had ever before online. Then, things started to show up at her house. Her mailbox filled up with empty formula cans, though she had no idea how anyone had found her address to send them to.

Shannon wondered if the strange missives were coming from haters online or people in her community. She grew worried. Did everyone in her neighborhood know about the formula thing? What about the other parents at her kids' school? Everywhere she looked she felt judged. More than ever, Shannon felt like the walls of Alpine were closing in on her.

Then, she said, Child and Family Services showed up at her house. Someone had called in a tip that the Bird children were in danger, and the agency needed to do a full investigation to clear the charges. Her kids had to be interviewed. Shannon was relieved when the officers seemed to be confused as to why they had been called to the Birds.

"You live in a seven-thousand-square-foot house," she said they told her. "Your kids are eating takeout sushi right now. Like, what are they talking about?"

While she can make little jokes about it occasionally, Shannon was extremely traumatized by the DCFS visit. Dallin, on the other hand, is so easygoing that she said he was never really concerned when DCFS came, calling the whole saga "ridiculous."

That's his attitude to most things Shannon posts online, including the formula saga. When I asked him if online criticism ever bothered him, he shook his head with a laugh. Even he doesn't really understand how he's able to not let it bother him.

"You know, I just don't care," he said.

Sure, he may wish she didn't post every single thing that comes into her head, but he long ago made his peace with the fact that he can't control what Shannon wants to do. He is capable of tuning out the opinions of strangers.

"You have to get to a point where, like, it's funny. It's funny to you," he told me. He launched into a parody of their haters: "The Birds are the dumbest people we've ever met. They're so annoying. We hate them. They suck. Their kids are embarrassed that they are their parents!"

Dallin laughed. His attitude is, bring me your worst. It's all part of the business.

"If you really, really care, then you can't do this," he said.

But Shannon is growing concerned about the impact the online perception of her could be having on their kids in real life after the incident. Hudson and Holland, her two oldest kids, were shaken, and it impacted their view of their mom's career, she said.

"Hudson and Holland are not fans of Instagram, especially after the whole thing . . . and they were very emotionally upset about that," she said.

Other family members were also upset. When Dallin's father saw the news report, he immediately got on a plane from New Jersey and showed up at the Birds' house. One day during his visit, he took a hammer to her pantry, which had insufficient storage space, and then spent weeks building a new, huge space for her to store food for the kids. Clearly, he felt that if Shannon had had more room to store baby formula, the incident wouldn't have happened. Shannon said he blamed himself, feeling like he hadn't been there enough for his son, Shannon, and the kids, even though that wasn't the case.

The incident is reverberating through Shannon's life. She feels like moms in her community have never stopped judging her since DCFS showed up at her door, almost as if their presence has marked her. She's now the mom who had Child and Family Services called, who was accused of something so horrible that the government had to get involved. Even those close to her throw it in her face. The day I arrived, someone she knew intimately had texted Shannon something rude and thrown in something about the incident, which devastated her. In the two years since it happened, Shannon has begun to suspect that the person who called DCFS on her wasn't some random person from the internet, but someone she actually knows. She may never know for sure. Regardless, it isn't the peanut gallery on the internet that is affecting her day-to-day life.

"It's still being used against me; I don't like that. They can use this against me," she said. She's worried that someone else may call DCFS on her, or that people in her community won't want to associate with her. She's worried people will judge her kids, or won't let their kids play with hers. She's worried that the stigma of being a "bad mom" is one she will never be able to fully wash off. "It's like, I don't feel safe."

Though it's not her primary concern, the online vitriol is impacting her career. The combination of the backlash and, likely, the beginning of the COVID-19 pandemic led to Shannon's influencer income cratering. Every time she did an ad, it would be attacked online by people who would inundate the brand with the insistence that Shannon was a bad person who didn't "deserve" to work as an influencer. Brands would get cold feet, saying, "Oh, you had cops called on you, so we can't work together anymore," Shannon said.

It happened over and over again. Take one victim of outrage

against her, a yogurt company she had partnered with in May 2021. She posted an ad of herself sitting on a snowy hill, eating yogurt. Chaos ensued. Shannon did an impression of the melee for me: "They're like, 'She doesn't eat yogurt for breakfast! This is bullshit!'" She doesn't think they cared if she eats yogurt or not. People just will attack her for anything, because they don't want her to be able to make money on the internet anymore.

The yogurt company was disgruntled and probably will never work with her again. Another bridge burned. And the cherry on top of the whole thing was that she really liked the photos she had posted for the ad because they were of a fun family outing to go skiing. When she had to delete the ad from her feed, she also deleted the photo from her family memories on Instagram. At a certain point, she wondered why she was even doing this. It was too stressful.

"Every time I do an assignment, I have anxiety," she said. Every time, she is just thinking, "Please, let me just keep this [post] up."

Talking about the impact on her career, Shannon grew more and more frustrated. These people are taking away her livelihood and everything she built. Apparently, it's not enough to have more than ninety thousand followers on Instagram and to be able to effectively sell products. Shannon doesn't deserve to make money as an influencer, people have decided, and therefore, her career should be torpedoed.

In the fall of 2020, she decided to try what she called a "science experiment." If these people were going to spend all their time ruining her career, didn't they deserve to compensate her for her lost wages? She was constantly being slandered, she said, and every company she worked with was getting tons of hate. She felt like she couldn't work anymore.

Before the 911 debacle led to a constant deluge of haters her way, Shannon was making about $13,000 a month from her influencer career and had about 13,000 people viewing each of her Instagram Stories. After the incident, she actually had more eyeballs on her than ever. Her follower count on Instagram spiked, going from just under 100,000 in December 2019 to about 104,000 in March 2020. Clearly, the haters didn't want to stop following her.

Shannon couldn't understand why people claimed they wanted her to leave the internet and never come back, yet so many continued to watch her stories and consume her content. They wanted her to continue to entertain them, but they wanted to make sure she couldn't be compensated for it, she came to believe. Shannon realized that if every person who watched her stories gave her a dollar, she would make up for all the lost revenue. She began to think that was only fair.

"You guys have taken away my income. You should pay to be here," she said.

So, she launched her experiment, to try and make a point. She had wanted a reconstruction of her boob job for a while, and it just so happened to cost around thirteen thousand dollars. She started a GoFundMe titled "Let's Fix Mom." Shannon wrote that after a tough year, she needed the surgery to recover from breastfeeding her five children. She also didn't want to spend Dallin's money to do it.

"This has been a year! Not just for me, but for everyone and I am more than sensitive to this. After my story went viral I was left to pick up the pieces, broken more than I started. I lost my income for almost a year now, gained more hate than any one person deserves in a lifetime," she wrote.

Predictably, the fundraiser got a huge backlash. People raged against her, calling her "selfish" and "disgusting." Women who claimed to be in treatment for breast cancer said they were enraged she would ask for help when people like them deserved breast reconstruction more. Soon after launching the page, she took it and the blog post announcing it down and said she would give any funds raised back to donors (because the page was removed, I'm unable to verify these claims or how much was donated).

Shannon told me that she wasn't actually trying to beg for donations just to be selfish. She wanted to see if people would compensate her for her lost wages. It's not just about the money, though. Shannon is, or at least was, proud of everything she has built for herself. By being self-sufficient and making her own money, she's setting a good example for her kids, and providing a better life for them. Recently, Holland has been curious about money and one day asked Shannon, "Mom, who pays for stuff?" It was cool for Shannon to tell Holland that she had her own job and her own money, not just her dad.

Now she doesn't know what she has.

"No one wanted to even touch my brand anymore, so I, like, shot myself," she lamented. "I went through a self-esteem crisis. I went through an identity crisis. I was like, 'What am I anymore?'"

WHILE SHANNON STRUGGLED to contain the fallout from perhaps being too authentic on social media, Caitlin began to chip away at the identity she had cultivated for so long on the internet: the perfect girl.

That's not to say that Caitlin thinks she is perfect. Far from it. She would categorize herself as an average person—living a blessed life, sure, but not a perfect one.

This perfection that the industry strove for, though, has been one of the damning criticisms of it in recent years. In many ways, the influencer industry has taken on the criticisms that have always dogged the fashion, beauty, and magazine sectors. These forms of media create an unrealistic image for young women to look up to, and according to many experts, they can cause more harm than good.

When I was growing up, the models that graced the magazines I read and that I stared at on billboards on the highway were incredibly thin, with almost no deviation. There was no movement to show "real" bodies in the media. Everyone was a size 0 and celebrated for it. Celebrities were shamed in the tabloids for supposedly gaining a pound or two. Undoubtedly this had an effect on my self-esteem and the way I saw myself, and my teenage girl peers. No one ever told me it was okay to be comfortable in my skin if I was over a size 2, that's for sure.

Social media has taken on these roles in our society, but is even more widespread.

"While influencers and celebrities have existed for millennia, they haven't held this kind of influence and power, with such an incredible lack of checks or balances," reporter Christianna Silva wrote for Mashable in October 2021. "They're marketers, inspirations, and an unfair comparison all rolled into one. And because it is relatively new, we're still coming to terms with the potential harms of this violently underregulated industry."

The issue came to a head in September 2021, when a *Wall Street Journal* report revealed that internal Meta data had found that Instagram was having a harmful effect on teenage girls. In fact, they found that "thirty-two percent of teen girls said that when they felt bad about their bodies, Instagram made them feel worse," the newspaper reported. Meta found that these problems were unique to Instagram because it has such an emphasis on image, aesthetics, and how you present yourself.

"The features that Instagram identifies as most harmful to teens appear to be at the platform's core," *The Wall Street Journal* reported. "The tendency to share only the best moments, a pressure to look perfect and an addictive product can send teens spiraling toward eating disorders, an unhealthy sense of their own bodies and depression, March 2020 internal research states. It warns that the Explore page, which serves users photos and videos curated by an algorithm, can send users deep into content that can be harmful."

Another study, reported by *The New York Times* in March 2022, was a little less dire, saying that while social media doesn't impact teenagers across the board in a negative way, its effects are felt. "Taken together with past work, the findings suggest that while most teenagers are not affected much by social media, a small subset could be significantly harmed by its effects. But it is impossible to predict the risks for an individual child," the newspaper reported.

Much of the responsibility for the studies of the effects on teenagers fell on the shoulders of Instagram, rather than individual influencers. While influencers can certainly be blamed for presenting a sanitized or unrealistic version of their lives if they are doing so, the researchers placed more blame on the addictive nature of social media apps and their algorithms, and the harmful rabbit holes it encourages young people to go down in order to increase their reliance on the apps.

However, the influencer industry has also been doing some soul-searching and self-correcting. When Caitlin started blogging in the early 2010s, perfection was the name of the game. But by 2014, perfection was falling out of fashion. Some influencers began to realize that by filtering their photos, dressing in flattering outfits, or even posing or standing the right way, they were presenting themselves in an unrealistic fashion. One of the earliest influencers to find success by proudly showing off their "flaws" was a little-known blogger named Rachel Hollis, who went mega-viral in 2015 by posting a photo of herself on Facebook celebrating her "honest" bikini body.

"I have stretch marks and I wear a bikini. I have a belly that's permanently flabby from carrying three giant babies and I wear a bikini. . . . Those marks prove that I was blessed enough to carry my babies and that flabby tummy means I worked hard to lose what weight I could. . . . They aren't scars ladies, they're stripes and you've earned them. Flaunt that body with pride!" (Hollis then turned her influencing career into a motivational speaking career, three bestselling books, and much, much more we can't even get into here, as much as I'd love to.)

Soon, influencer after influencer was going viral for similar posts (when I started working at *BuzzFeed News* in late 2014, I swear we wrote a post a day about one of these Instagrammers). Once a few influencers were praised in viral magazine articles and on talk shows for showing themselves unfiltered, many more followed, creating microtrends of their own. Countless fitness bloggers would show before-and-after shots of themselves posing and flexing, versus not. In the first photo, the influencer would show herself looking taut and muscular, but the second would show how her abs, chest, and arms really looked. "Instagram versus reality," many would write, which also would go on to become a meme.

Caitlin had known for a while the wind was shifting. She had to be able to open herself up more to her followers, even if that meant deconstructing the walls she had built to protect herself against criticism all those years ago when she was devastated by the reaction to her breakup with Chris.

"I feel like people are really connecting to those bloggers and the whole curated perfect thing that people don't want to see anymore," she said. "And it's hard for me to shift my mindset because . . . I've been doing this for ten years, and that's what I've always done."

Caitlin began to let people more and more into her inner world, and her followers responded positively. In a "Birthday Q&A" post in 2018, Caitlin casually announced that she had decided to get rhinoplasty about a year prior, writing that she hadn't mentioned it on the blog but "it might be obvious to some of you."

"I didn't post or talk about it then because I didn't want anyone to think I was promoting plastic surgery—it was a deeply personal decision, and one that I made for myself and for my confidence (after years and years of thinking about it and researching it)," she wrote.

In the comments, her readers lauded her for her candor. "This is probably the most honest blogpost I've ever read!" wrote one. "Thank you for your honesty about your nose! I'm considering mine, and yours looks so great. You're just amazing and I'm so happy to follow you," wrote another. "These answers are incredibly honest and that's SO refreshing . . . especially your answer about plastic surgery! I love you a little more now that we got a deeper insight!" said another.

In 2019, Caitlin decided to open up to her followers and the world about her long struggle with anxiety and depression. Her decision to share led to one of the biggest boosts to the visibility she'd had thus far, a feature on *People* magazine's website. In the article, titled "Caitlin Covington Says Social Media Fame Caused Crippling Anxiety," she opened up about what she described as her long battle with mental health issues that stemmed from the pressure she felt in her influencer career.

"I had never experienced anxiety or panic attacks until I started blogging," she told the magazine. "All of the sudden I had all of these eyes on me, and they were starting to pick apart my personal life. People can be so cruel on the internet. . . . I could be doing something simple and then all of a sudden I would have a panic attack."

She also dispelled the notion that her Instagram was a true depiction of her real life, explaining that she saw her content as aspirational, not realistic.

"In a lot of cases it's art, and you want your art to be beautiful," she said. "But that is in no way a reflection of that person's real life and the struggles that they're going through."

By January 2021, Caitlin had been thinking about this Instagram versus real life conundrum a lot. She would soon be undergoing a huge life change, as she was about to give birth to her first child, daughter Kennedy. Recently, she had been wondering if this life change could lead to another big shift—in her business mindset.

Caitlin was planning on sharing as much of her postpartum journey as she could, such as her body shortly after giving birth, and just being real with people. "I want motherhood to be a chance for people to see the more raw side of me," she said.

She had wanted to open up more for a while. She knew she risked becoming a dinosaur if she clung to her old ideas of curated beauty and perfection, and she didn't want to get left behind. "I'm having to kind of break my walls back down and be more real with people because I don't think bloggers will survive if you don't show more of yourself in the industry how it is today," she told me.

But there was more that went into her decision. It wasn't just about business. She genuinely admired the other mom bloggers she saw who showed their babies' spit-up all over their tops or got real about motherhood, and she could see that their audiences valued that honesty too.

Caitlin decided that she didn't want anyone to compare themselves to her, for them to think that she was "a perfect mother with a perfect baby."

It was a scary proposition, though, because of how much she had been burned in the past. And it wasn't like doing this when she was about to become a mother was going to make her critics go easy on her. There's nothing more that people love than shaming or questioning a mother for every decision she makes. From the moment she gets pregnant she is scrutinized. That happens even if you aren't sharing your life with more than a million people every day.

Caitlin had already experienced her fair share of mom-shaming before her baby had even entered the world. She had posted about wondering if her baby would come early because she had been one centimeter dilated at a doctor's appointment. Her messages were flooded with people telling her that she was an idiot for thinking that she would have her baby soon, and she obviously didn't know what she was talking about. She was just making conversation, and besides, if she didn't know what she was talking about, wasn't she just a clueless first-time mom like everyone else?

If she posted a photo of herself with a Starbucks cup, she got a flood of crazy messages. It didn't matter that it is perfectly safe for pregnant women to drink coffee in small doses, or that the cup might not even contain coffee. Caitlin would cope with these waves of criticism by just ignoring her DMs or having her assistant filter them for her. Still, it had been hard for her to balance the conflicting feelings of wanting to share but also being tired of being criticized.

"I'm really nervous to become a mother and receive messages about how I'm mothering because, in that situation, you can't make everyone happy," she admitted to me eight days before she gave birth. "It's going to be hard to navigate just knowing what people are and just knowing how this pregnancy has been and how outspoken they are and how they have an opinion on everything. And then becoming a mother and having a real human child, they're going to be even more outspoken."

It's a tricky balance, but I do believe it is an important one. Caitlin isn't going to do much good by presenting a sanitized version of motherhood, but by sharing her reality, she could actually help other parents. I have found influencers such as Caitlin to be invaluable resources when facing big life events like pregnancy, and I know my friends have too.

Since I started following bloggers as a young adult, I feel like they have guided me through many big life events. Take buying a house, for example. My husband and I purchased our condo in Brooklyn in 2020, in the middle of the COVID-19 craziness, on a whim and with little real understanding of what we were doing. What got me through the stress of that process was reading blogs from influencers I follow who had purchased homes recently as well. I read more than a dozen blog posts, hopping from one blogger to another as they shared how their family had made such a monumental purchase.

For many women, pregnancy is one of the biggest of these big life events that they are searching for information on. Many of them find comfort in the experiences of bloggers and influencers like Caitlin, who are sharing their journeys to motherhood in real time. They are interested to see how she adjusts to motherhood, but they also want someone to relate to. If an influencer shares how she is struggling to conceive and how she is coping, that can be comforting to other women who have been dealing with the same issues. When a blogger shares her baby registry, it can cut through the noise for the overwhelmed first-time mom who is drowning in stroller reviews and experiencing decision fatigue. When Caitlin shares her birth story, it can help other moms prepare for and maybe even feel more confident going into their own deliveries. Because they trust the influencer they follow, that influencer's word carries weight.

This is a valuable service that many women like me have almost come to rely on influencers for. I brought this up with Caitlin, telling her I believe that sharing is providing a valuable service. Several of my friends, I said, had recently gotten pregnant, and had been telling me they had been following a bunch of new influencers to prepare for pregnancy. Caitlin could relate to this experience. She follows people for the exact same reason and even buys products via affiliate links from people she admires on Instagram because of the connection she feels with them.

"It is really cool, the trust you have in that person and just the connection you feel with them, even though you've never met them," she said. Being one of those people can be incredibly rewarding, but as we said our goodbyes, I wondered how Caitlin's new "raw" personality as a mother would actually be in practice.

On January 16, 2021, Caitlin gave birth to Kennedy. "How do you put in words the most meaningful moment of your life? I was, and still am, afraid that I can't do it justice. I think you other mamas will agree with me, there is simply nothing in the world like meeting your baby for the first time," she later wrote on her blog.

In her birth photos on her blog and Instagram, Caitlin admittedly looks great (she always looks great). She also looks incredibly happy. The first photo she posted of herself and her daughter shows her beaming, eyes shut and a look of genuine happiness on her face. To her side stands Chris, full of genuine emotion.

Caitlin did indeed keep sharing many "unfiltered" images of herself as a mother, and her followers were loving it.

In one, posted about a week after Kennedy's birth, Caitlin holds her daughter while dressed in leggings and a sports bra, looking tired but happy. She has no makeup on, her hair is in a bun, and her stomach still swells with the remainder of her baby bump.

"No makeup, dirty hair, messy house, crying baby and yes that is a diaper I'm wearing," she wrote. "It has simultaneously been one of the happiest and hardest weeks of my life. I could not imagine loving anything or anyone as fiercely as I love my daughter—at moments it physically feels like my heart will burst. But being a new mother is HARD. I have cried every day this week, battled sleep deprivation and struggled to get the hang of breastfeeding. My house is a mess, I haven't had time to wash my hair and my daily wardrobe consists of either sweatpants or pajamas. Despite the challenges, I wouldn't change it for the world."

The post got 125,000 likes, nearing as many likes as the first photo she had shared of Kennedy just one week prior. It seemed that her followers were responding to Caitlin's new style. "This is one of your best performing posts and one of the most raw/relatable posts you've ever made," wrote one follower. "Please keep sharing this type of content. There is nothing more inspiring than a new mama! You are a superwoman!"

As Kennedy grew over the next several months, Caitlin continued to share photos of her reality as a new mom. In one, she sleeps holding her daughter in her arms ("super unflattering . . . but this is what I look like 95% of the time . . . just trying to keep it real!").

For Caitlin, everything had changed. For one, motherhood, while beautiful and amazing, had been a lot more challenging than she had anticipated. She had initially told her manager, Kirstin, that she would probably take two weeks off, and then would be able to get back into work slowly. After all, how hard could it be? Newborns sleep a lot, and Caitlin figured she could begin to pick her work up again pretty easily.

Yeah, no.

"I was really just blown away just by how hard it is," she admitted. "Your postpartum body, the pain that you're in after birth, all the emotions, the hormones, just having this baby that you love more than anything you've ever loved in your whole life, I was just totally blindsided by everything. It's been a huge adjustment."

Caitlin ended up taking more time off of her blog and long-form scheduled posts than she'd expected. But she surprised herself by actually spontaneously posting *more* than she had planned on her Instagram Stories in the initial weeks of Kennedy's life. While she was in the thick of early motherhood, she felt compelled to share what she was really thinking and feeling with her followers. Maybe it's because she realized she had been so unprepared. She had thought she was ready for all the changes coming into her life, but she had been blown away by how difficult some aspects of raising a newborn, like sleep deprivation, could be.

Caitlin realized that she had been consuming an idealized version of what the postpartum period would look like herself, and she wondered why no one had given it to her straight. She jokingly texted her friends, asking them why they hadn't told her it would be this hard. Why hadn't they been more real with her? "Y'all make it look so easy," she wrote. Even on social media, she said, she had followed other influencers who never let on how difficult it could be.

"It's interesting to see some bloggers, like, just almost act like their life hasn't changed dramatically. Every day it's just posting about outfits and I barely even see their baby," she said. "I'm sure they're doing it on purpose. They want to keep it just fashion or whatever. But I just couldn't do that. It just doesn't feel natural to me to just continue focusing on fashion."

Maybe the old Caitlin, the one who focused more on aspirational aesthetics, would have felt compelled to keep up that facade. But as the new person she was becoming as a mother, the way she used to do things didn't seem to make sense to her anymore. How could she post about deals and fashion when so much was going on behind the scenes?

Perhaps it was this feeling of being blindsided that led Caitlin to share more about her experience than she had anticipated. When Kennedy was three months old, she wrote a candid blog post about how difficult the transition had been for her. So many things had changed, she wrote, that she struggled to remember what life used to be like. Her anxiety was flaring, and she found herself googling constantly and worrying about Kennedy. She loved her daughter more than she had loved anyone, but sometimes she grieved the loss of her old, carefree life with Chris.

"It's so strange, because in many ways I still feel like 'me,' but in many ways I do not," she wrote. "Sometimes I don't even know who I am anymore. . . . To sum up motherhood: it's constantly feeling hundreds of intense, conflicting emotions all at the same time. . . . I feel all of these emotions, all the time, and they change moment to moment. And you know what? *It's okay.* I'm figuring it out."

When she published the post, Caitlin was shocked by the immediate response of appreciation and love from her followers. She doesn't remember ever getting so much feedback on a post. Women were pouring their hearts out to her, telling her they had felt bad in the past watching influencers who had made it seem so easy.

Caitlin realized that by holding back, she was missing out on this genuine connection with her readers.

"I think I could honestly be better at sharing more moments," she said of the past. "But motherhood has definitely brought out that side naturally in a way that I couldn't have expected or didn't plan for. It was just very natural. And the feedback from people has been incredible. It makes me want to share more real life, not just the picture-perfect moments."

The support has also given her the confidence to share despite any negative feedback she may receive. Before she had given birth, Caitlin had signed a deal to design a clothing line with the Instagram-famous online boutique Pink Lily. As part of the contract, she had agreed to travel to Mexico to shoot photos to promote the line when Kennedy was six weeks old ("Again, it shows how I underestimated motherhood and giving birth and the whole newborn stage," she said). By the time the trip rolled around, she didn't want to break her contract, even though she realized it would be difficult to leave her baby so soon.

Predictably, she got a ton of hate for going on the trip. People sent her messages accusing her of being a bad mom, telling her she had abandoned her baby just to "lay on a beach all day." The messages really hurt her feelings, but she just repeated a mantra to herself: "I know I'm a good mom. So what anyone else says about me, it doesn't matter."

After talking herself down, she decided to do something about it. She figured people may not realize that she was working the entire time she was in Mexico. They shot for four days straight, from sunrise to sunset. Caitlin wanted to tell her audience that she was doing this to provide for her child. She went on her Instagram Stories to set the record straight, saying that she is a working mother and would not stand for disrespect or judgment for the way she chose to juggle having her career and a baby. "I'm not going to tolerate these hateful messages," she recalled telling her audience. "This is a work trip, not a vacation. I have to do my job like any other working mom."

Surprisingly, some of her followers seemed to understand. They even thanked her for speaking out against the stigma that working mothers face. But what made Caitlin even more confident in her decision to go on the trip was how she felt afterward. She realized that she had missed being her old self, and working on the shoot had given her the perspective and confidence she needed after her rough entry into motherhood.

"We were in this beautiful place and we were creating content all day long and it just was invigorating," she said. "I walked away from that feeling so inspired and ready to get back to motherhood and ready to be with my daughter again."

Caitlin was doing motherhood her way, and she finally wanted to share it. While the haters still bothered her, her confidence in her decisions was helping her navigate tough times in both her career and her personal life in her own way, and she felt like she was even improving both in the process.

In many ways, she felt more confident in her career than ever. After years of trusting her own instincts, from quitting her job to taking her first ad deals to navigating life online, Caitlin decided to continue to trust her gut. She knew what was best for her business, and for her followers.

"I have just tried to share and do what feels natural to me and I've grown and adapted all along the way," she said. "I feel like I'm constantly reinventing myself."

SHANNON HAS two big things to tell me.

It's January 2021, and the Birds are going on their first big trip since the COVID-19 pandemic began. But it's not just that they are going to Florida for ten days, it's who they are going with: Papa John, a.k.a. John H. Schnatter, the founder of the pizza chain. Schnatter resigned in disgrace at the end of 2017 over his comments about NFL players kneeling in support of Black Lives Matter and has continued to face backlash for his extremely racist comments that pop up every so often. Shannon said she met Papa John (she only calls him that) hobnobbing in Park City. After he sent her photos of himself on his yacht, he offered to take her family to Disneyworld via private jet.

Her second bit of news is even bigger. Shannon has been interviewing for *The Real Housewives of Salt Lake City*, which had been about to finish airing its first season when we talked. She tells me she was considered for the first season of the show as well but wasn't sure if she would be able to fully participate because she was pregnant at the time. They circled back when considering possible replacements for season two. She even had been set up with one of the show's stars, Jen Shah, who was indicted in 2021 in a fraud scheme, for what she called a "sleepover" to test how they gelled.

As a viewer, I think Shannon seems perfect for the show. She thinks so too, and so does Dallin, she said, because he knows she has no filter and is, in his words, "emotionally unstable enough that she would definitely jump at this opportunity." Shannon cackled.

Plus, working on a TV show could be her escape. Shannon is tired of being a mom influencer, and of the corner she's boxed herself into. She's shrewd and knows being on TV could take her influencer business to the next level.

"This would be a great springboard into something bigger," she tells me. She's already scheming how she could turn a potential TV opportunity into a business, which she says a producer told her was the difference between Real Housewives who make "$50,000 versus a million" dollars a year. She's considering a "mom brand," TBD.

"It would be awesome for them to do all of it, for me to live my life and they film it; that would be easier for me," she says.

While Shannon didn't end up getting the gig, she didn't stop thinking about leaving influencing behind for good. The girl the haters love to hate wanted to say goodbye.

But could she?

7

In the video posted on Instagram in January 2021, Mirna looks radiant. With a huge smile, she begins to speak.

Her calm demeanor was welcome in the chaos that had erupted in the world and online. It had been only two days since a group of right-wing protestors stormed the US Capitol in defiance of the 2020 election results, and Mirna was giving an online talk, which she later posted to her social media, to the Green Mountain Club, a conservation group in Vermont, and she had no interest in pretending the events of January 6 had not occurred. Right up top, she told the audience that if they hoped she wouldn't mention it, they would be mistaken.

She continued, "As a Black person who frequents outdoor spaces for my personal enjoyment, for work, for my physical and mental health and well-being, I am very keenly aware of my existence and my body in spaces that are often seen as white spaces. The great outdoors in the American imagination does not exist apart from the erasure of Indigenous communities and the theft of their homesteads, the forbidding of Black folk from certain outdoor spaces both de jure and de facto, the continued media representation of one kind of body, one kind of experience, one kind of outdoors."

Mirna hoped that even if some in the audience felt uncomfortable, they would stay anyway and listen to her perspective. Ultimately they might learn from it. That is an ethos that Mirna lives by, and one of the main things she tries to impart to the world: nothing and no space is apolitical.

Every day, by choosing the life she loves, Mirna is inherently making a political statement. She is a Black plus-size woman who is happiest when outdoors—running, biking, and walking on the trails—in spaces that, as she said, are usually stereotyped as white spaces. Everything she does in her career is to change that narrative.

Mirna taught in boarding schools for nearly twenty years and still considers herself an educator. When she was a teacher and frequently was the only Black person or one of a few in the white spaces of elite boarding schools, she often took on the role of diversity, equity, and inclusion coordinator at many of the schools where she worked. At her last school, she served as the director of equity and inclusion for three years, developing a curriculum to work with both the staff and the students on racial justice issues.

In 2020, after the murders of George Floyd and Ahmaud Arbery made headlines, Mirna began to wonder if she could use the experience she had gained as a diversity educator in schools to impact a larger audience. She was already talking about Black Lives Matter and diversity frequently on her Instagram, but she thought she could maybe do more. This was partly because, she said, people she knew—white people, that is—were asking her to help them.

Well, she thought, sure, I can help you, but I'm going to do this my way. In July 2020, she made an announcement on Facebook. She would be hosting a virtual "interactive workshop," which she called "Introduction to Identity, Social Justice and Antiracism."

"Maybe you don't know that I was also an educator for 18 years up until 2018, and a large part of what I did was Diversity, Equity, and Inclusion education," she wrote in the announcement. "These days, I've been called to step back into this work because our country is suffering right now. Lots of folks have asked WHAT CAN I DO? WHERE DO I START? HOW CAN I HELP ERADICATE RACISM? . . . I believe that when you know better, you do better. Knowing and examining your identity will help you honor, love, and protect those with differing identities. Learning about social justice and what 'the work' is will help clear a path for you!"

Mirna set a price of two hundred dollars for the course, which would be a little under five hours. She opened up fifty spots, which sold out in two hours. Just like that, she had a whole new career.

Her business consultant told her she needed to open up the course to more people. Mirna was hesitant, saying she didn't know if she wanted to be responsible for so many people. "Who cares?" she recalled her consultant saying. "Lean into it." So Mirna did, opening the course up to 180 people. Since then, she has been teaching her courses to corporations, including her longtime sponsor Merrell, to great success. It's not just lucrative, but extremely fulfilling. But it takes its toll on her mentally and can be exhausting.

Still, the requests keep coming. Mirna had always received many requests to share her unique story, but since the Black Lives Matter protests in 2020, they have been unending. She had never publicly announced that she moved to Vermont, but soon the local media called. Everyone, it seemed, was eager for her to represent Vermont. At one point, she was on the cover of two local magazines at the same time. One article, on a local news website called VTSports.com, dubbed her "Vermont's New Face of Fitness."

"You know there are other Black people here, right?" Mirna joked about her reaction to the press. In one interview with *Seven Days* magazine, she explained how interesting the flood of requests has been.

"Black is hot right now, and I'm in the outdoor space, so there's a lot of, 'Can we talk to you about being a BIPOC in the outdoors? What are your suggestions?'" she said. "And it's great to see that happening right now, but I can't fix anyone's problems."

The year 2020 saw a huge, long-overdue shift in the influencer industry. Influencers, and those who support them, were forced to reckon with the fact that they had a major diversity problem.

Beyond the actions of individual influencers, though, the Black Lives Matter movement of 2020 prodded the industry to take a hard look at how Black influencers were treated among their ranks. It wasn't pretty. Black influencers had fewer followers than white influencers. They were paid less than white influencers. They were given fewer opportunities than white influencers. And the statistics were stunning.

In December 2021, the Influencer League, a group founded in 2019 to create opportunities for Black and POC content creators, published the results of a study they did with MSL, a public relations firm, to examine the racial pay gap in the influencer industry. Their findings, published a year and a half after the Black Lives Matter movement caught fire in 2020, revealed stunning inequality. The racial pay gap, the study found, between Black influencers and their white counterparts was 35 percent, and between white influencers and BIPOC influencers was 29 percent.

"These are stark numbers by any measure," D'Anthony Jackson of MSL said in a statement. "Just compare the 35% gap between white and Black influencers to the pay gaps in other industries—education 8%, business and financial 16%, construction 19%, media sports and entertainment 16%. The gap this study uncovered in influencer marketing vastly overshadows the gaps in any other industry."

It's not that Black influencers with as many followers as white counterparts were getting paid less, though. Rather, a Black influencer would face an uphill battle to reach the same size as a white one. In fact, the study found that just 23 percent of Black influencers, versus 41 percent of white influencers, fell into what they deemed the "macro-influencer tier," of more than 50,000 followers. These "macro" influencers make on average, they said, $100,000 or more a year, contributing to the huge racial pay gap.

Seventy-seven percent of Black influencers fall into what they deemed the "nano and micro-influencer tiers," under 50,000 followers. Having fewer followers doesn't just mean that these influencers make less money, although the pay gap between this cohort and the next highest tier is vast, with these influencers making just $27,000 a year. Micro and nano influencers are more likely to be offered free products instead of money in exchange for a campaign, as brands can argue they offer less value. This is despite the fact that numerous digital marketing agencies have pointed out the value of smaller influencers in particular, who tend to have highly engaged audiences who trust them and appreciate their more intimate connection with their audience. (In fact, the social media marketing platform Later found in a 2021 study that the smaller an influencer's follower count, the higher their engagement rate was on average.) They may have trouble finding representation with an agent or manager, who can do the complex calculations and negotiations required to ensure they are paid what they are worth. Left on their own to navigate brand deals in an industry where opacity is the name of the game, many Black influencers often get the short end of the stick.

"Pay opacity disadvantages BIPOC influencers with unequal access to information or professional advice both in pricing themselves and in negotiations," the MSL study concluded. "This was most pronounced with Black influencers, 45% of whom cited 'managing the financial process' as their most challenging pain point of working with agencies and brands versus 27% of white influencers."

Black influencers also face a higher penalty for speaking out about social justice issues than white influencers, especially Black issues. Of the Black influencers surveyed, the majority of them said they were passionate about racial justice, and 59 percent felt they were punished for speaking out on their platforms. "Juxtaposed with the outpouring of support by brands for the Black Lives Matter movement and the many commitments companies have recently made to racial justice, this was particularly striking," the study authors noted.

Fellow influencers and industry commentators can be passionate about leveling the playing field, but brands ultimately decide whom they partner with, and whom they pay fairly. Still, influencer marketing entities like MSL and the Influencer League are working to help bridge the gap, by creating more transparency around influencer pay and making public databases that are available to all.

Managers like Kirstin are also trying to change brands' perceptions of who is worth more or less. She has been trying to dispel the notion that follower count should be the metric brands are considering above all else, because Black creators have a much harder time reaching certain metrics. She has noticed that some of her Black clients will get "stuck" at around half a million followers, while white creators seem to sail to over a million quickly.

It's been disgusting, she said, watching the uphill battle Black creators have had to climb to build their platforms, versus their white counterparts.

"The amount of Black creators that even have a million followers on Instagram in comparison to the amount of white creators is so disproportionate it's insane," she said. "It will make you sick to your stomach."

It's particularly baffling, she said, because many of her Black clients bring so much to the table. Their engagement is "out of this world," their content is "incredible," and they offer a huge advantage to brands who want to work with creators of color. So the idea that they should be paid less than a white creator simply because of follower metrics is one she said she is working to change.

"They have an insane, engaged audience, and it just doesn't make sense to me," she said. "They should be getting those higher rates because if they were a white creator, they would be at a million followers."

The new spotlight on the industry's racial equity issues created opportunities as well as stressors for Black influencers. In the months following George Floyd's death, many influencers began to speak out about the clear diversity issues in their industry. Some white influencers promised to push back on brands that had no influencers of color as part of their marketing plan. Others began to share their favorite influencers of color, shouting them out and urging their followers to follow them as well. Like Mirna, these Black influencers were suddenly inundated with attention they hadn't really experienced before. Many, frankly, had whiplash.

One of them was Ayana Lage, a lifestyle influencer and blogger from Tampa, Florida. Ayana had started her blog, XO, Ayana, in 2017 after working in journalism and marketing for a few years and feeling like she had hit a creative rut.

Ayana loved blogging and content creation, and she gained a respectable following of about thirteen thousand on Instagram. She even started to get brand deals and began to realize that if she played her cards right, she could make decent money from this side gig. In 2019, she made around twenty thousand dollars from her Instagram business—not life-changing money, but enough to make her try and pursue influencing more seriously.

However, Ayana struggled to really take off and felt like growing by even a few hundred followers was a constant struggle. She stressed about her modest growth rate, analyzing each piece of content and reading all she could about the industry. She didn't understand how some people seemed to grow so easily when she felt stuck at thirteen thousand.

Then, George Floyd was murdered on Memorial Day 2020. Ayana was horrified and wanted to do something to stand up for

Black people everywhere, but at the time she was seven months pregnant and didn't feel comfortable going to a local protest. So, she decided to use the platform she had: Instagram. She posted a series of Instagram Stories about why she believed it was so important for everyone to show public support for the Black Lives Matter movement in the wake of Floyd's death.

"The first thing that happens is that you signal to people who look like me that you are on our side. . . . The second thing . . . is that when you publicly say this country was built on systemic racism, Black Lives Matter, I understand why Black people are angry, you will get your white friends and your white relatives . . . responding to your posts . . . and they will be angry . . . and you get to educate them instead of people like me having to do the work," she said.

After she posted on her Instagram Stories, a few friends reached out and asked her to make the videos into an IGTV, so they could share on their own pages. Ayana did, and the response blew her away.

She became the talk of Instagram overnight. Huge white bloggers whom she had never heard from before began to reach out and share her posts. They began to encourage their followers to follow Ayana, and they did. In just a week, she gained approximately thirty thousand new followers. Her video reached nearly one million views, and hundreds of DMs filled her inbox.

Part of her was thrilled to finally have a bigger platform. Part of her felt totally overwhelmed and sad. Finally, she was getting what she had always wanted and worked so hard for. After grinding away for so many years, she felt she deserved the success she was now seeing, and she felt proud of herself.

But this well-deserved success had come from such an ugly moment. But then, that wasn't *her* fault. She hadn't posted what she did expecting this to happen. She wasn't trying to be exploitative; she was just trying to be a Black woman on the internet standing up for her people. But then . . .

It also was, frankly, weird. White influencers were sharing her videos and telling people to follow her, but why had they never noticed her before? Did she have value to them only now, when they felt compelled to share Black influencers? And so many white people were following her, but did they expect her to be their teacher on racial justice? Her DMs were unrelenting. Some people were nice. Some asked her so many questions she felt overwhelmed. Others sent her racist comments or tried to start arguments. She couldn't help everyone understand the moment, but she felt guilty if she wasn't at least trying.

Ayana was providing thousands of well-meaning but ultimately kind of clueless white people with the education they needed, but for free. Because while her follower count had shot up, she wasn't making any more money than she had been the week before.

But there was hope. Now that she had the platform, maybe she could finally be compensated in an equitable way, and brands would finally notice her and see her value.

Two years after Ayana's big viral moment, some of her wishes had come true. Soon after her video went viral, big brands began reaching out like crazy. In the first six to eight months after June 2020, the number of requests she got for brand partnerships was overwhelming. She landed deals with big brands like Old Navy and launched a podcast, *Asked by Ayana*, on which she has interviewed stars like Mandy Moore (and one non-star, yours truly). She now has about forty-eight thousand Instagram followers and is repped by an agency, Shine Talent Group.

Considering that Ayana had given birth to her daughter in August and taken a few months of maternity leave, the amount of money she was making and her interest from brands was even more staggering. December 2020 was her biggest month ever, and she thinks if she hadn't taken maternity leave she likely would have made even more. In 2021, she made six figures off her influencing career—around $115,000—for the first time.

"That was pretty wild to me because I don't feel like I'm expending much more effort than I was back in the day," she told me. "I feel like I work less and care less."

Ayana began to think that maybe this was her life now and that she would always make this much. Over time, though, things started to level off. The brand deals bonanza began to die down, and 2022 started slowly. Ayana did not continue to gain followers as she had been, plateauing at around fifty thousand. At first, it was hard for her to watch the fervor around her account die down. In the first six months since she went viral, she lost about five thousand of her new followers, which she took personally. She wondered if they had followed her only because they saw her "snappy" social justice videos and thought that's why they should follow her account. But those videos weren't really what her account was ever about. At the end of the day, she is a lifestyle blogger, who just happens to be Black. She thinks it is important to talk about social justice, but she doesn't want that to be her entire account.

All in all, Ayana has become more confident as an influencer since before she went viral. She'd spent so many years doubting herself and obsessing over every piece of content, and getting upset when she wasn't growing the way she wanted to. But when she posted something that was authentically her, it resonated. She wasn't thinking about an algorithm; she was just speaking from the heart. So these days, she takes it all a lot less seriously. She posts the way she wants, and if brands and followers respond, that's great. But she knows what kind of influencer she wants to be.

"I'm just grateful," she said. "I have this following and brands pay me good money and I'm having fun with it. So, yeah, I'm a little bit less hard on myself."

Overall, Ayana thinks the influencer industry has a long way to go in fixing its serious equity issues. But she has seen some improvements. For one, she feels followers are way more accepting of and even expect influencers to share their political views. She also thinks followers are holding brands accountable, which, although it is a small thing, is slowly helping make changes in the industry.

"I think that brands are realizing that there is an optics issue and if they underpay a Black influencer or if they do a brand trip and it's all white people, people are going to say something," she said.

AS THE NEWS has gotten more dramatic and more polarizing in recent years, the issue of how to address current events has become a huge one for influencers like Caitlin. They had never before really been expected to weigh in on the news of the day, although some would. Most influencers, if they acknowledged anything at all, would share memes or photos that circulated through social media every time there was some sort of tragedy in the news ("pray for Paris," "Boston Strong"). But they didn't feel like they had to be arbiters of complicated sociopolitical events, and for the most part, their followers didn't expect that of them.

For a long time, this kept up with the precedents set in traditional entertainment. Reality shows, like *The Bachelor*, for example, were in their heyday during the Bush administration and the war on terror, but show host Chris Harrison and contestants never felt compelled to share their viewpoints on the war, and the viewers never demanded or expected it.

But the norms have changed, and quickly. One reason is that millennial and Gen Z consumers don't just want to consume, they want to consume ethically. That translates to the products they buy, as well as whom they buy from, and whose opinions they value.

In 2020, though, the desire for ethical influencer consumption reached a crescendo. Many people simply couldn't associate with people who weren't handling the coronavirus pandemic the way they were, and they were determined to suss out who on the internet was on "their side" or not. As people who were both in the public eye and accessible to the average person, influencers were the perfect vehicle for this anxiety. In online forums and comments sections, followers began to pick apart how their favorite influencers were handling the pandemic and watching them for a hint of a mistake.

Influencers grapple with these types of quandaries all the time. Despite the lack of respect they get in mainstream society, they are always expected to set a good example and to do the right thing. At some point or another, every influencer is going to mess up. That's just life. There are small slipups, like accidentally saying the wrong thing or posting at the wrong time. But even small incidents can blow up to be hugely consequential for their reputations and brands.

Take, for example, a scandal that rocked a baby-sleep expert and parenting influencer named Cara Dumaplin. Cara, better known as "Taking Cara Babies," had become huge in recent years for her popular infant and toddler sleep courses, and easy-to-follow tips and infographics she shared with her more than one million Instagram followers. Around the day of President Joe Biden's inauguration, someone decided to look Cara up on the Federal Election Commission's website and found she had donated to the Trump campaign and a pro-Trump super PAC thirty-six times between 2016 and 2019. People soon began to call for Cara to be canceled over her support of Trump, while another sizable contingent began to support her, saying her politics had nothing to do with her work. (Cara confirmed to me at the time that she had made the donations.)

Situations like Cara's are pretty black-and-white. If a follower doesn't want to support an influencer after finding out about their politics, they can unfollow them easily. Where it gets more complex is when influencers are expected to react to every major news event in real time and ensure they aren't sharing disinformation or an uninformed opinion. In many ways, in the politically charged atmosphere of 2020, influencers faced controversy no matter what they did. If they didn't say anything about a news event, they were accused of not caring about it. If they shared something deemed half-hearted, they were decried as phonies. If they shared an infographic or information in real time that was proven later to be false, they were even worse off.

For Caitlin, navigating these minefields grew more and more difficult by the day, as she grew more and more pregnant through the end of 2020 and into 2021. When we spoke soon after the insurrection, she was less than two weeks from delivering, and the events of the week had taken their toll.

Caitlin had realized that she needed to speak out about her political beliefs earlier in 2020. She knows what people think when they see her. She's white, pretty, privileged, and Southern. She's even made being "Southern" a key part of her brand. Everyone assumes she's a Republican.

In 2019, a nineteen-year-old college student tweeted out a photo she had found of Caitlin and her friend, Emily Gemma, from a blogging trip they had been on. In the photo, both women have bouncy barrel curls, huge blanket scarves, matching skinny jeans and ankle boots, and matching totes.

"Hot Girl Summer is coming to an end, get ready for Christian Girl Autumn," the student tweeted.

The college student told my colleague at *BuzzFeed News* that she had found the photo by googling "cute church outfits" and "all scarf outfits." The "Christian girl autumn" meme persists to this day, as a shorthand for "basic white girl," although the photo of Caitlin and Emily is sometimes supplanted by a group photo of a bunch of other influencers on a fall blogging trip.

Caitlin has always had a good sense of humor about the whole thing. After the photo first went viral, she tweeted, "If all of Twitter is gonna make fun of my fall photos, at least pick some good ones! Super proud of these. For the record, I do like pumpkin spice lattes. Cheers!" To accompany the tweet, she shared some fall photos she preferred of herself.

When people post about Christian girl autumn, they usually assume the women in the photo are racist, homophobic, or both. Caitlin and Emily took pains to distance themselves from the assumptions. Caitlin responded to many questions on her Twitter account, answering that she wasn't a Republican and that she supported gay rights. When the student who created the meme came out as trans in 2020 and created a GoFundMe to pay for her transition, Caitlin donated to the fund and tweeted out the link, encouraging others to follow her lead, generating a new round of press.

As the election rolled around, though, she grew concerned that her followers didn't truly see who she was, or were assuming things about her. She spoke out, posting that she was supporting the Biden ticket. It was uncomfortable to be so public about her personal beliefs, but it felt right.

"I was really nervous to say something, but I just didn't want people to think I voted for Trump," she said. "I just feel like he's such a monster. I just want people to know where my heart stands."

It wasn't just that Caitlin, on a personal level, wanted to stand up for what she thought was right. She also knew that her role had changed. She could no longer be silent about her political beliefs and continue to post like nothing was going on. She had witnessed a turning point, and influencers were now expected to speak up about their social values.

There are also the business ramifications of staying silent to consider. She knows people are ready to unfollow her if they think she doesn't align with them politically.

"That's so important right now that [followers] don't want to purchase from an influencer if they know they don't align with their values," she said. "They don't want to follow you. They don't want to purchase from you. That really, really matters to people."

With more than one million Instagram followers, Caitlin knows that she is speaking to an audience with a huge diversity of opinions. There's no way she can please everyone, and no matter what she says, someone is going to get pissed off. Standing up for what she believes in inevitably means that she is going to lose income.

"I don't want to alienate people and make a lot of people upset and unfollow me and not support me," she said. "But I also want to stand up for what I believe in."

Ultimately, Caitlin has to be at peace with potentially losing revenue in order to do what she thinks is right. Speaking up about Trump, gay rights, and Black Lives Matter comes easily. She is firm on where she stands, and won't apologize for it. This has also been good for her reputation, with people commenting on Reddit they have been "pleasantly surprised" by her views.

What's trickier is navigating the never-ending news cycle, to respond in the moment to huge events like the insurrection, the Supreme Court nominees, and crises abroad. Now some followers expect her to respond to everything, and to do so perfectly.

"We are expected to watch the news and have one second to digest it and then make a statement," she said. "And I think that's so unfair. I think we should be allowed at least twenty-four hours to really process what's going on ourselves and then eloquently say something about it. Instead, we're expected just on the fly to have an immediate opinion and to have it be the right opinion and to have it be the politically correct opinion. And that's so hard because we're human and we're not always going to say the right thing. And even if we have the best intentions, it's always going to offend someone."

Kirstin, her manager, said that she feels like the burdens

placed on influencers like Caitlin are unfair. Of course, she said, people with huge platforms should use that platform for good and speak out about the things they believe in. But, she said, no one should be going to influencers for news, or comprehensive social justice platforms.

"What we should do on this platform is acknowledge that this is not right and we don't stand for this, but to expect a full synopsis summary on the intricacies of it . . . expecting influencers to do that is not fair, in my opinion."

Take the January 6 insurrection, for example. Caitlin was overwhelmed and horrified. She was also thirty-nine weeks pregnant, and her emotions had been going haywire. She had barely been able to watch the news because her anger and emotions would overwhelm her. She wasn't doing well. Adding to her stress was her job. Should she say something on Instagram? If so, what? What if her followers attacked her? What if she accidentally said the wrong thing? Her anxiety overwhelmed her, to the point she felt paralyzed.

"I just can barely keep it together. I've been so angry and upset by everything going on," she said. "It's just been hard to process myself. I can't process it. And then put out a message to everyone this week, I can't do it."

She decided to just go radio silent. She didn't feel like it was appropriate to post her usual content, so she simply took some time off from Instagram.

"Everyone was messaging me like, 'Are you having your baby?' No, I'm just having an emotional breakdown," she said.

Some people may have thought this was the right choice. Others didn't. Caitlin got messages telling her they were unfollowing her because she didn't post anything about January 6. But she was at peace with her decision. At certain points, her mental health had to come before her work.

"If you're going to have longevity in the industry, you have to be able to know when you're not in a good headspace to post," she said. "You have to take breaks."

I have watched with interest as people's feelings toward influencers and social responsibility have changed over the years. People have been easing the pressure on influencers just a little bit in terms of expecting them to respond to every news event.

As I write this, the world is watching in horror as Russian troops invade Ukraine. I was surprised to see that, at least in online snark forums, the tide seemed to be shifting on what responsibility influencers had. Commenters on Reddit seemed to understand influencers may not have the perfect thing to say and were okay with that.

"Can we all agree to refrain from the tired influencer xyz didn't bring up the Russia/Ukraine issue today?" one person wrote. "One can be angry and scared about the situation without making a public statement about it (especially when one is not educated on the subject). I find it upsetting and I am scared for the people of Ukraine but I don't plan on making a statement at work today about it, nor would it be appropriate for me to do so."

At the end of the day, Caitlin knows that it wasn't her the angry commenters were mad at. They were mad at the state of the world, at the state of the country. Since the pandemic had begun, she'd noticed how much angrier everyone online seemed to be, a sentiment many influencers have shared with me.

Caitlin thinks she's an accessible target. She was easily reached via DM if anyone needed to blow off steam.

"I think influencers just the past year have been an easy target, almost a punching bag, and when something's going on in the world, it's like, 'Let's look at influencers. What are they doing?'" she said.

This issue is a complex one, and no one really has the answers. Ayana has struggled to answer it too. She had encouraged people to speak out on social media about Black Lives Matter, but is it fair for followers to yell at her in the DMs if she isn't posting enough about Ukraine? She doesn't want to get it wrong, or not share the information in a responsible way.

"How do you decide what issue is, I guess, worthy of you? . . . There's so many news accounts on Instagram. I'm not one of them," she said.

8

Shannon's older children had slowly begun to chafe against the demands of her career. Holland especially had come to resent being a character on Shannon's blog and Instagram, and as she grew, she became less and less willing to participate. Instead of smiling for the camera and hamming it up for videos like some kids on social media, Holland would frown and roll her eyes. She and Hudson set the tone for the younger kids, so they all would rebel when it came time for Shannon to create her content.

One day in late 2019, Shannon needed to film some content for a Christmas partnership with a school uniform company called French Toast. Six-year-old Holland and eight-year-old Hudson led one of their typical rebellions. None of the kids were cooperating. Shannon became more and more anxious trying to get them to behave and smile for the photos. All she wanted was for her kids to look perfect, so the brand would be pleased with the end result. She had gathered the kids in her bedroom, against a white wall, and they just wouldn't participate. She grew more and more frustrated. The company was paying well, and it would take just a few minutes to get the shot if the kids would just play ball. The money she brought in paid for their vacations, their clothes, and their hobbies. Why did they have to make it so hard for her?

Finally, she lost it. She started screaming at her kids, telling them they wouldn't be able to leave the room until they "got it right." She even tried to hide the cast on one of her son's legs because it didn't match the aesthetic. The pressure to please the client was too much.

Once she had cooled off, she had an epiphany. Maybe, she thought, she needed to just be done with this. What was she doing to her children?

"I was like, 'What is this childhood?'" she said. "Is this worth it?"

Shortly after, the 911 incident in early 2020 derailed Shannon's career. Since then, she has had a lot of time to reflect on how her career as a blogger has affected her as a mother, and her relationship with her kids. She is not actively seeking out new campaigns, though she does some if they come to her. She also has slowed down posting about her kids, choosing to post more about herself instead. She even changed her avatar on Instagram from one of her and her kids to one of just herself.

These changes are due to the toxicity her brand has been tainted with since the 911 incident, and it's been harder for her to find work. But they also reflect Shannon's choice to slowly transition herself and her kids away from being a "family of influencers" and give them more space and privacy. For so many years, her career dictated her children's lives. In the beginning, it was easier because they didn't have any real opinions about what they wore or what they did. But now in hindsight, Shannon realizes that sometimes her career determined her kids' childhood, not the other way around.

Take Halloween. Shannon's first free perk ever from being a blogger was gifted Halloween costumes for Hudson, sent to her in exchange for a blog post. For years after that, the Bird family's Halloween costumes were dictated by sponsors. The kids would wear the costumes they got for free from whatever partnership or deal Shannon had managed to secure that year, Shannon would get a brand deal, and everyone was happy.

As the kids grew, they started to want to assert themselves and their own creativity. They didn't want to wear whatever sponsored costume came their way; they wanted to pick out their own. For a few years, Shannon resisted. Frankly, these costumes were free and were helping to pay the bills. Her kids could suck it up for the good of the family. Recently, she has been reconsidering. She now lets her kids have their own costumes for their activities with friends and trick-or-treating, and makes them wear the sponsored ones only for obligatory content on her page. It's a small step, but it's just one of many ways she's reconsidering the way that her motherhood and her career have merged.

Shannon isn't entirely sure why her kids, the oldest two especially, have chafed so hard against becoming characters on her blog. She sees so many families in which the kids seem delighted to participate in photo shoots and videos with no furtive eye rolls like Holland especially tends to throw in. She doesn't get how some parents can get their kids to happily participate in not just photos for blogs and Instagram but also mediums like YouTube videos, which take much more cooperation, time, and energy.

Dallin believes that the only way families are able to be on YouTube to that level is if the kids are also invested in the content. Otherwise, there's no way it would work. "The kids, I think, like it enough for the perks and the money and like the benefits enough that they're willing to go along with it at that level," he speculated.

Shannon has tried to get her kids more personally invested in making content, in hopes it will both teach them about money and responsibility and get them more excited to participate. After Holland and Hudson became able to consent to their own campaigns, Shannon began asking them if they wanted to do them and letting them keep the money for themselves. Now both of them collect their earnings, usually a couple hundred dollars a pop, if they consent to doing their own campaigns. The Birds treat it as an allowance for their kids, giving them a portion to buy what they want.

It works, sometimes. Holland recently did a campaign with a toy company, in which she posed on her bed surrounded by a bunch of free dolls. She got to pocket that cash, which Shannon said she was pleased with. (Despite this, Holland told me she doesn't like doing sponcon for her mom's Instagram.)

As we discussed Holland's influencer cash, Shannon's oldest son, Hudson, chimed in to say that once his sister made $350 . . . for just one photo! Doesn't that make him want to do it too? He said no, but Shannon told me sometimes he does complain that he wants to get as many jobs as his sister does.

"Well, I say, 'Hudson, if you want jobs, you need to be in my pictures,'" Shannon said. Otherwise, she tells him, how will companies even know he exists?

Hudson was quiet. He didn't seem convinced that taking photos was worth it. Shannon said the kids doing their own jobs gets her only so far. They still push back against the content she is getting paid for.

Even when it's fun, the kids seem to know that they are working, not having a spontaneous family moment. It doesn't make a lot of sense to Shannon. She tries to get her kids excited about the opportunities they are given through her work, experiences she would have killed for as a kid. But the kids realize they are on the clock and resent it. When Shannon got the opportunity to take the kids on a sponsored trip to an amusement park, which came with six hundred dollars to spend there, they weren't enthused. All they would have to do is take a few pictures, and do maybe ten Instagram Stories.

"They would almost rather not even go sometimes . . . they'll pitch a fit about it," she said.

Shannon places some blame on herself. She would bark at her kids to fix their hair, for example, because people on the internet criticized the way her kids' hair looked.

"It's because I get so intense," she said. "I started being a stage mommy about it."

How to responsibly feature kids on the internet is one of the thorniest issues that the influencer industry is having to consider as it evolves and matures. Shannon's dilemma is not unique. The kids who have grown up being filmed and discussed on their parents' social media accounts are growing up, and are slowly beginning to have agency. They are beginning to ask, "What rights do I have to my own image? Aren't I owed some of these profits? Can I say no?"

When I first started following blogs, so-called mommy blogs like Shannon's were central to the industry. The radical act of unabashedly recording motherhood for public consumption has changed the lives of countless women across the US and has upended our viewpoints of how mothers are supposed to behave.

The prominence and prevalence of these types of blogs probably explains why I, as a single woman in my early twenties, started reading them in my spare time. In fact, I read mommy blogs more than fashion blogs.

Is it that I have always been interested in motherhood, as many young girls are, getting excited when a character on a television show becomes pregnant? Am I some kind of sick voyeur, constantly craving the intimacy of this content? Was I just really, really bored and lonely, and watching women who looked like me make happy families for themselves gave me comfort? Maybe all of it. But it wasn't just me. In 2009, around the peak of the mommy blog obsession, a study found that twenty-three million women were engaging with blogs, by either reading and commenting on them or writing their own, every week.

The fact that young children were at the heart of these accounts has always been a source of shame—for me as a reader, and for the industry as a whole. The subjects of this content are real children, like the Bird kids, who are starting to realize their entire childhoods, or at least large swaths of it, have been documented on the internet. And that means they are at the mercy of the internet's judgment.

Shannon used to post "crazy" things Holland would do, like cutting her own hair, because she genuinely thought they were funny family moments. But to anonymous posters on the internet, they were fodder to tear her daughter down.

"Of all those kids Holland grates on me the most," wrote one person in May 2021 on GOMI. "Her attitude is awful and I can't tell if it's because she dislikes her mother, if it's because she's feral, or is she ostracized social bc of that significant speech impediment (my girls had lots of friends/social time in grade school and I see no evidence). IDK."

Shannon's Instagram posts also began to affect her daughter IRL. Holland told Shannon girls at school would bring things up to her, so Shannon stopped posting things that could embarrass her daughter, or that showed her in anything but a happy or neutral light.

"People don't call her crazy anymore," Shannon said.

In my own contemplation about kids of influencers, I had figured that the number one issue parents may be concerned about is safety. Surely, posting an image of your child to thousands, in Shannon's case, or potentially millions, in the case of bigger influencers, can be scary. What if someone gets fixated on them, and approaches them on the street?

Shannon's perceived lack of concern for her family's privacy is also a frequent topic among her snarkers on Reddit, who frequently insist that by sharing their lives, she is setting her kids up to be kidnapped.

In 2020, one person wrote, "24 hours ago Shannon was talking about a creepy man who was too interested in the children. Today, she's leaving out bread crumbs by tagging the location of their favorite running trail and letting us know they travel half of the trail just her alone with 5 kids. I don't even know if she understands what she's doing, or if she does and just doesn't care."

The Birds, though, aren't overly concerned about posting their children on the internet. The way the world is now, they think, everyone is posting their families, most for anyone to see. It's not taking a risk, it's just the new way we all live.

"That's just reality," Dallin said. "You could be a nobody and take pictures of your kids at Disney World." He doesn't see much difference between what Shannon does and the photos average parents without a huge following share online publicly every day.

It is a little jarring, though, to actually realize how many people could recognize influencer kids in the wild. A few years ago, I was in Trader Joe's in Chelsea in Manhattan, waiting in one of its typically long lines. I realized right ahead of me were some bona fide Instagram celebrity children: Tao and Ren, the kids of Eva Chen, the Instagram director of fashion partnerships who is also a mega Instagram influencer herself. Eva wasn't even there—they were with their father—but I knew them immediately.

Ren and Tao have it easy, though. They are side characters to their mom's account, not the focus, and they don't spend that much time, relatively, on camera. The violation of children's privacy on the internet can get much, much worse.

Of course, like anything, there's a scale. Some influencers don't show their kids at all, and some discuss their kids but don't expose everything about their lives. Some influencers, though, film their kids seemingly all day, every day, exposing the good, the bad, and the ugly.

Currently, the children of content creators have no legally protected rights, but that may be about to change. Incidents over the past few years are beginning to demonstrate how kid-centered content can be a slippery slope into darkness, exploitation, and abuse. To best demonstrate these issues, we must examine a different corner of the social media universe: YouTube.

Myka Stauffer and her husband, James, were family vloggers. It's a genre that you're familiar with if you spend any time on the platform, but if you have never seen them before, the videos are rather strange. They literally just film themselves all day, not doing much besides going to the grocery store, doing chores, or walking around their neighborhood. It's bizarre, but very popular, especially among young children, who spend hours and hours on YouTube on average every week.

Myka and James had been on YouTube for about two years when, in July 2016, they announced they were planning to adopt a young boy with special needs from China. Over the next year, their impending adoption was a huge part of their channel. They hosted fundraisers, answered frequently asked questions, and revealed all the ins and outs of the adoption process.

In October 2017, the Stauffers welcomed their son, Huxley, to their channel with a video titled "Huxley's EMOTIONAL Adoption VIDEO!! GOTCHA DAY China Adoption," which they said was dedicated to "all of the orphans around the world." It got more than 5.5 million views, more than any other video on their channel before or since. Over the next few years, Myka posted updates on how Huxley was adapting to the family, sharing that the boy had been diagnosed as "having a stroke in utero, has level 3 autism, and sensory processing disorder." As Huxley adjusted to the US, Myka's fame and prominence on YouTube grew. She positioned herself as an adoption expert and advocate in outlets like *Parade*, and partnered with brands like Glossier, Good American, Fabletics, and Ibotta. Her channel grew to more than seven hundred thousand subscribers. But slowly, over time, Huxley began to be featured less frequently, before disappearing from the channel altogether.

In May 2020, James and Myka posted a video titled "an update on our family." In it, they revealed that they had decided to give Huxley to another family, who they said was better equipped to deal with his special needs. Their attorney told me that the couple was "forced to make a difficult decision, but it is in fact, the right and loving thing to do for this child."

The story I wrote about the Stauffers became the most-read piece I have ever published for *BuzzFeed*, attracting seven million views. I received countless emails, and Instagram DMs from horrified readers, begging me to investigate further (I ended up publishing several more stories on the Stauffers, including one in which local authorities confirmed Huxley was safe in a happy home).

People tended to be angry about two main things. First, they felt Huxley had been exploited by the Stauffers on their channel. The second was that some people were horrified that so much of Huxley's life had been used on a monetized channel. People online began to call for the monetized videos to be removed. A Change.org petition on the matter was signed by more than 150,000 people.

On the one hand, the Stauffer story is an extreme example of how the overexposure of a child on social media can lead to bad outcomes. Huxley, a special needs child, had been adopted into a family where he was conscripted into contributing to the family income without any personal compensation or right to privacy and then cast aside when he no longer could work within the family unit. It was horrifying, and a clear-cut case of the dangers of monetizing your family and using children for paid advertisements.

But most influencers don't give away their kids to new families, and many don't share as many private and personal details of their kids as the Stauffers did when sharing Huxley's adoption journey and their experiences dealing with his special needs.

However, the dynamics at work apply to anyone who makes a profit from content featuring their children. These children, commenters agreed, deserved a right to privacy, to not have every detail of their lives shared with strangers on the internet. There needed to be some sort of rule book in place to standardize what parents can and cannot share about their kids without their consent. And if children are working in a family business making ad revenue, they deserve a cut of the profits.

I began to wonder if the Stauffer case was going to be the tipping point that caused the general public to actually start to seriously examine these two issues, and I was not alone. A few days after I published my original story, I received an email from a woman named Rossana Burgos, the matriarch of the "Eh Bee Family," a popular YouTube family channel with more than ten million subscribers (they recently changed their name to "The Bee Family.") If you don't know them, google their name plus "gif," and you'll recognize immediately that their family is one of the most well-known reactions for "celebration" on the internet.

Rossana wanted to thank me for writing about the fight to protect kids on YouTube, which she told me she had been privately battling behind the scenes for years. She has been trying to get YouTube and other platforms to "protect children who are exploited every day for views," she wrote, but she was barely making progress. She was tired of watching families involve their kids in inappropriate or dangerous stunts on the platform, and continue to profit off the exploits.

When Rossana reached out to me, these concerns were at the forefront of her mind, as her family had just dealt with a horrifying incident. A person on Facebook had created a "fan page" for her then-fourteen-year-old daughter, and cropped an innocuous still image from one of their videos to look inappropriate. It had taken the family more than a week, and countless pleas to Facebook, she told me, to get the photo taken down.

Rossana's role as a wildly successful parenting content creator gives her a unique vantage point into the world of those who make money off content featuring their kids. What she has seen has horrified her.

"Things can very easily get out of control in this business," she said. "When you incorporate money with kids, it can be a very, very dangerous equation. . . . When you see some of the things these families are doing in the name of making money, it's really dangerous and I don't think that we will see the effects for another ten, fifteen years," until now-children are grown.

Rossana is bothered by the amount of time children are spending being filmed, and thinks the parents who are detailing all the minutiae of their children's lives online don't really understand the consequences of what they are doing. Women who share their children's potty-training journey, for example, think they are helping other moms. They aren't thinking about the possible consequences for their kids.

"I just don't think they have the awareness to really understand what they are doing, the foresight, that emotional intelligence, to realize," she said. "Yeah, it's cool when they are two or three; it won't be so cool when they're sixteen and there's videos of them on a potty."

Family vloggers on YouTube and parenting influencers on Instagram like Shannon are not exactly the same, and the potential for exploitation on YouTube is much higher because so many family vloggers film their kids for several hours a day. Their audiences are also very different. Children and young people are the prime audiences for family channels like hers, Rossana said, and their demographics on YouTube skew younger than on other platforms.

"It's this perfect storm, where there's people that shouldn't have a platform, who shouldn't have a camera, who should not be allowed to put out content. You have unsupervised children who are watching this stuff, who are thinking that this is normal, and then you have the kids who are in the videos, who don't realize the situations they are being put in," she said.

In contrast, she said, the audience for their family Instagram accounts skews older. If the primary audience for YouTube family vlogs is other children and teens, the primary audience for many parenting influencers is their peers, and other women looking for advice and community on their own parenting experience.

But the risks are still there, and the backlash coming from extreme examples of toxic YouTube culture could trickle down to affect influencers on Instagram. If the exploitation of kids on YouTube leads to new laws or protections for children, those laws will likely apply to Instagram influencers as well.

There have been some tepid efforts to regulate and protect kids through legislation, but none have gained any real traction. Some have suggested that child performers on the internet be regulated under guidelines similar to those that regulate how child actors are treated.

In the early days of Hollywood, no regulations existed to protect the emotional or financial rights of children who appeared on-screen or onstage. That changed in 1939 when California enacted what is most commonly referred to as the "Coogan Law." According to SAG-AFTRA, the labor union representing film performers, the law is named for former child star Jackie Coogan. He entered the film industry as a child in 1919, soon becoming a star in several films with Charlie Chaplin. When Coogan turned twenty-one, however, he discovered all the money he had made was gone. His parents had complete control of his earnings and had apparently squandered them. Coogan sued his mother and former manager, and the law that helped protect kids avoid this same fate bore his name.

Almost a century later, SAG-AFTRA and others have worked to strengthen the original law to better protect the assets of child actors. In 2000, California law changed to ensure that any income minors made from the entertainment industry was their property, and did not belong to their parents. In California and many other states, parents and guardians are required to set aside 15 percent of child actors' gross earnings in a trust for their future use. In addition, most states have laws that regulate child actors' employment, with some of the strictest, like California's, restricting how many hours a child can work and other provisions.

None of these laws, though, apply to children making money on the internet. That's a problem, writes Marina Masterson in a 2020 piece on "kidfluencers" for the *University of Pennsylvania Law Review*. "Because kidfluencers have no legal right to these earnings or safe working conditions, the risk of exploitation is extreme and immediate," Masterson writes.

Masterson, however, acknowledges the issue is complicated to fix. After all, even regulating child actors has been a struggle, with a "patchwork" of state laws regulating the industry rather than a federal mandate, she says. Regulating kidfluencers, who are most of the time being filmed or photographed in their own homes by their parents, is even more complicated.

As Masterson writes, "certain common child actor regulations, like those involving work permits and workplace conditions, are difficult, if not impossible, to impose on kidfluencers." This is due to the nature of how the content is produced and filmed. There's no set, no working hours, and no script. Rather, the filming is spontaneous, in their own home, generally without a set schedule.

For example, some states dictate how many hours a child can be on set in a typical film production, which is pretty easy to enforce. But it's much trickier to enforce a work hour limit, Masterson writes, when the "set" is the child's own home. "Even if the state set an hour limit that these children can work, the only way to enforce that rule would be to monitor the families within their own homes, which would be an overstep by the state," she writes.

Thus, trying to make Coogan Law protections apply to kid influencers would be "largely unworkable in the fast-paced social media context, which is generally confined to the family unit," she writes.

"Financial protection is immediately possible through Coogan Laws, but regulating the content production itself presents new and challenging questions that require states to consider the specific needs of the social media industry."

In 2018, according to *The Hollywood Reporter*, legislators in California introduced a bill that would add "social media advertising" to a list of employment that constitutes child labor in an attempt to regulate the influencer industry. "Under this 'kidfluencer' bill, minors working in the digital sphere would have to obtain a work permit and follow measures similar to those required by the Coogan Law," the magazine reported.

However, the effort didn't achieve what advocates had hoped. Although a version of the bill was passed, it was "diluted significantly," as *The Hollywood Reporter* put it. Masterson writes that this was because by the time the law was signed, "any mention of social media was removed." This attempt and failure, Masterson says, proves that social media influencing needs its own set of laws to properly regulate it. "Current child actor laws should not simply be expanded to include social media influencers, but instead, tailored legislation is needed," she writes.

"All child performers face some common harms—missing school, losing privacy, and exerting labor at an age where they have less personal agency. But the social media context introduces an additional host of harms, most notably the total lack of financial protection and the health risks associated with interactive media and extreme loss of privacy. . . . The risk of exploitation in the social media context is double-barreled—children face exploitation both by their parents and by the companies that sponsor them," she writes.

Ultimately, Masterson says she believes that at the very least, Coogan-style financial protections should be enacted to protect kidfluencers. But she concedes that the other issues are complicated, recommending lawmakers "continue to research and refine the appropriateness of other regulations."

As of now, though, she writes, kids and parents are mostly on their own to regulate themselves: "Children spend hours per day producing high-valued content at the direction of their parents with no financial or personal protection besides the good will of their parents."

While American legislatures aren't taking steps yet to protect kids, some other countries are. In 2020, French lawmakers passed a piece of legislation aimed at protecting underage social media stars. The law addressed the two main issues facing these kids: protecting their privacy, and making sure they are retaining a portion of the money they have earned from their work online. The law would apply only to children who spend "significant" time working online, ensuring they will get access to their earnings as adults and regulating how long they are able to work.

The law also included an interesting provision. Children would have the "right to be forgotten," *Le Monde* reported. If a child decided as an adult they no longer wanted to be internet famous, platforms would be legally required to erase content they were featured in.

As of now, though, nothing like that exists in the US. This leaves the responsibility of protecting a child's privacy, and the management of any funds they earn, to the, as Masterson says, "good will" of their parents.

Rossana said she and her husband are doing their best to ensure both. Each member of the family owns 25 percent of their family channels, and they have set aside their son's and daughter's shares of the profits in trust funds. When they come of age,

Rossana hopes, they can pay for their education with the money, or use it in some other way to help them start their lives.

Shannon and Dallin also are trying to set up a good foundation for their kids through what they have earned as child influencers. Shannon seeks to make sure that the kids are getting to actually use the perks they are earning through campaigns. In 2021, for example, Hudson got a cell phone after doing a campaign with Gabb, a start-up that makes "safe" cell phones for kids. The kids' money also goes into investment accounts, which the Birds hope they will appreciate once they are older and can choose how to spend their funds.

They agree that the industry needs to be better regulated, and their kids are not putting in even close to the hours that a family of YouTuber kids is. But Dallin, like Masterson, pointed out that actually regulating the industry would be extremely complex. How exactly would the government regulate the Birds taking a photo every week or so? And how would they draw the line between work and pleasure? If the Birds went to an amusement park as part of a campaign, for instance, is that work or play? It would be tricky to navigate because so much of the content being produced is not simply work or organic family life, it is both. That's what makes it so compelling, but also so complicated.

Rossana thinks that children can be protected online both financially and emotionally, but it would require more than just the government or the platforms to step in. It would take all of us collectively working for change. In addition to legal protections, she envisions a kind of "governing board" of mental health experts and teachers regulating the industry, ensuring the kidfluencers are emotionally healthy, are getting a real education (no unregulated homeschooling), and their earnings are safe. Everyone would need to be on board, she said. The platforms, the government, the brands, and the audience all need to step up and work for change.

Those are some potential solutions to the kids-on-the-internet issue on a macro level. But what does life as an influencer actually do to a family unit?

Since she has stopped taking as many campaigns, Shannon has started to reflect on how her own anxiety and stress about building her career have affected the way her kids feel about her. It's a hard topic to talk about, but she is extremely candid with me about it. Lately, she has been wondering if her career as a mommy blogger has impacted how her kids, especially Holland, perceive her. She's beginning to think it has, and to wonder if that has contributed to Holland's resistance to appearing in campaigns.

"I think I almost created it with having a blog," she said. She wonders if when her kids were with her in their early childhood, they began to feel like they were on the clock, and began to associate time with her as work, not pleasure.

Dallin agrees. He thinks that maybe the kids felt like their time with their mom was for her benefit, not theirs. They wanted their mom to be investing only in them, and they resented anything else.

"The kids perceive that you are getting them to do things to make *you* look good," Dallin suggested to Shannon. Kids like it, he said, when they are being supported in their own hobbies and interests. "When they are doing things for the blog, they see it as supporting you," he told Shannon.

So, they have begun to rebel against what they think is an artificial part of their life. When they see their mom making content, they don't want to be a part of it.

"They say, 'This is staged! This is fake!'" Shannon said. It leaves her in a tight spot. "It's hard. What do I say? Like, do it? This is my job. My kids are my job."

As Shannon begins to wind down her mommy influencer content, at least to some degree, Caitlin is just beginning to grapple with what it means to post Kennedy on her page.

It may be the most pivotal business decision of her career: how to protect her child and maintain her business. And there's no one who can answer that question for her. There's no governing board, as Rossana envisions, looking over her shoulder, providing a fact sheet for her to refer to for best practices. There are no regulations to guide her. There's no rule book. Actually, there are no rules. As Masterson said, whether her child is protected will depend solely on Caitlin's goodwill. She and Chris have to decide everything for themselves and hope for the best. All they have is each other, and their guts.

On the one hand, she and Chris believe that Kennedy's privacy is important. While Caitlin has always had control over her own career and what she posts, she also recognizes that Chris should have equal say over what information about his daughter is shared with the world.

"It's hard because there's, like, a line that we have to walk between people," she said. "While they're dying to know what the baby looks like, and what classes we're taking, and how we're going to sleep train, and all that kind of stuff. And I want to share that with them. But I also want to protect our daughter as well."

In fact, it feels natural to post about Kennedy. Caitlin had thought she would be warier of sharing her daughter with her followers, but if anything, she feels compelled to share her.

"I can't imagine just not sharing her," Caitlin said. "I'm just thinking about her all day long. I'm thinking about all the baby products we're using and motherhood. And so it just seems natural to share that."

It's a tricky balance. Caitlin has come to realize how much she values sharing her part of the motherhood experience with her readers, and how much they value her candor in turn. But Kennedy is inevitably a part of that journey as well. How can Caitlin transparently speak to her readers, while also making sure she is doing the best for her daughter?

Caitlin's Kennedy content, while frequent, is pretty benign. In the first year of Kennedy's life, Caitlin wrote only three blog posts entirely centering around her. Two were about her birth, and one was the post in which Caitlin opened up about her postpartum experience. Otherwise, she writes a lot on her blog about products she and Kennedy like and Kennedy's fashion, but she doesn't reveal a lot of personal details about her daughter.

Caitlin is happy to share photos of Kennedy. She thinks Kennedy brings joy to people, and she doesn't share anything controversial.

But what happens when Kennedy starts not wanting her picture taken, or doesn't want to be in her mom's campaigns? Caitlin's not exactly sure. Her philosophy on sharing her daughter so far has been to take it day by day, but she assumes eventually she will need to reconsider how much she shares. And one day, Kennedy herself will have an opinion about her presence on her mom's page.

"I'll definitely reevaluate, when she has an opinion and when she's a little older. . . . It's just different when they start to develop a personality and can talk," she said, adding, "I don't know. I want to take it day by day, I guess."

9

Shannon has Instagram. She has her blog. But there's another place where she has a substantial following: Poshmark.

Shannon doesn't just make money from her brand deals or from the savings of not having to buy clothes or home goods. If she doesn't like or want something she gets, she can always sell it.

On Poshmark alone, Shannon has more than 10,000 followers and has listed nearly 700 items. She can sell almost anything she gets, from her kids' Halloween costumes ($30 each) to a Mima Xari stroller (which retails for $1,600 but sold for $400). The kids' section of her Poshmark closet reads like a who's who of trendy baby clothing, accessories, and more: Janie and Jack, H&M, the Gap, aden + anais, Líllébaby, Restoration Hardware Baby and Child. On each listing reads the same label: sold, sold, sold.

Her Poshmark side hustle also gives her another title: job creator. For years, Shannon employed a girl she knew to help her sell things she didn't use. Shannon piled all the boxes of free stuff she didn't want in a corner of her house, where her resale assistant would spend her days organizing and selling it. In exchange for her efforts, she took 15 percent of the profits—enough money to make it her full-time job.

Like Shannon, most influencers employ or have employed staff—not just publicists and managers like Margaux and Kirstin, but assistants who keep it running behind the scenes. In this way, the influencer industry is not only making thousands of women content creators, managers, and agents wealthy, but it's also creating jobs for an entire class of influencer support staff, who are almost entirely women and usually either moms or young people who appreciate the flexibility.

Mirna not only has Margaux to work with her on her brand deals but she also has an assistant, Kimberly, to handle her administration. Kimberly is the daughter of a friend of Mirna's from her time living in the Riverdale section of the Bronx, and does things like answer Mirna's emails and run her schedule. Shannon had someone helping her organize and negotiate brand deals at one point as well.

Caitlin has observed some of her peers in the industry who have hired more than a dozen assistants on the payroll at once to take over almost every aspect of their day-to-day work, from answering DMs to writing their Instagram captions to helping them come up with content. She speculated one influencer has five assistants whom she passes off as herself in try-on photos and videos cropped to show only their bodies, from the neck down (you can tell because they have different body types, she said).

That's not Caitlin's style, though. For many years, her control-freak side was hesitant to cede any aspect of her business to someone else. She just didn't think anyone else could do it as well as she could.

"I just think the business only works if it's me," she said.

Even now, after she's hired her assistant, Chelsea, for three days a week, and her brother as a photographer, Caitlin struggles sometimes to cede control. She still plans her own content and answers her own DMs, except when she's getting hate and asks Chelsea to screen them for her. But at the end of the day, she needs her assistant. Caitlin's favorite parts of her job—being creative, styling her outfits, planning her content, and photo shoots—take up the smallest amount of her time because she has so many other things she needs to do.

Most of Caitlin's workday is spent strategizing about upcoming content. She plans out her blog posts, her Instagram Stories and Reels, and her posts. She considers how she'll approach holidays and major shopping events like the Nordstrom Anniversary Sale, which is basically the Super Bowl of blogging. The sale, which has been a mainstay of the department store since its inception in the 1960s, became a staple of many influencers' income through a mix of commissions made directly from the store's own affiliate program and those from other affiliate programs, like RewardStyle. For much of her career, Caitlin made about 75 percent of her annual RewardStyle commission in the month of the sale alone.

Caitlin's content doesn't just happen. It all takes a lot more effort behind the scenes than she thinks people believe. She is very hands-on, even editing all of her photos herself, instead of letting her brother, who has worked for her since 2018, do it. She doesn't think she could ever hire another photographer for this reason, as she doubts another professional would be down for her to edit their work. Even taking a simple Instagram photo can require hours of planning. She can't just stand in the corner and point and shoot.

"The planning it takes to maintain that aesthetic is really hard," she said. "We research a lot of places for photo shoots and you have to plan the colors of the outfits. And it's, like, so much work that people, just, their minds would be blown, I think." Caitlin is the fashion director of her own magazine, juggling many roles at once.

So having someone like Chelsea to take care of the simpler tasks has been a game changer. On a typical day, she arrives at ten a.m. and begins unboxing all the things they may need as Caitlin takes an hour or so to get ready for the camera and answers emails and DMs. Chelsea also helps manage Caitlin's content creation calendar and schedule. She usually leaves around two p.m., when Caitlin's brother arrives to shoot photos for future campaigns or content, which usually lasts the rest of Caitlin's workday, until five. Once she and Chris eat dinner and have family time with Kennedy, Caitlin starts working again. She edits photos, answers emails, and makes her editorial calendar, usually until about midnight. And these days, she's working less than she used to.

"Before I had my daughter before I got married, I would work 80-hour work weeks," she said. "It's way more balanced now, but for a long time, for years, I really grinded it. I would stay up till 2:00 a.m. working. And I loved it because when you have your own business, it's like your baby."

The tasks that Caitlin describes seem so simple but actually take substantial effort behind the scenes.

Take one thing that seems extremely easy: answering direct messages on Instagram. I truly did not understand how time-consuming DMs could be until I started my own Instagram page for work. I don't have a huge number of followers compared to Caitlin, Mirna, or Shannon—about 35,000 as I write this—but I am consistently blown away at how many DMs I get, and have gotten, even when I had a third of the followers I have now. And the thing is, I want to answer them!

But it is incredible how much time it takes. I probably spend at least thirty minutes to an hour a day during my workweek responding to DMs, sometimes more if I post something that gets a lot of responses. When you have more than a million followers, as Caitlin does, trying to answer them all can be a losing battle. There's just no way.

"I don't answer all of them, but it hurts me because then

people will be like, 'You used to answer my DMs, and then you never answer me anymore,'" she said. "It breaks my heart; I feel so bad that I can't answer."

She tries to be responsive in other ways. If she gets five hundred DMs asking her for a link to a sweater, she will then post the link on her stories (this is why influencers saying, "A lot of you have been asking . . ." has become a meme). She also tries to answer as many as she can in her downtime, like when she is feeding Kennedy.

Not surprisingly, this level of work can lead to burnout. When I polled influencers, one of the biggest stresses they cited was feeling as if they had to constantly hustle. When your life is your work, the lines become incredibly blurred between the two. One described the stress of needing to "have a constant content flow—it is always hanging over your head." Several described feeling "guilty" for turning off for the night or a weekend, and feeling like they can never take a break.

"There's no separation between my work life and my personal life," one said. "It's hard to 'turn off' when every single thing you do—from the second you wake up and eat breakfast, until you go to sleep at night—is considered content."

And while nearly all of them had at least one assistant or contractor they worked with regularly, they still ran a lot of their businesses themselves.

"I'm the CEO, the assistant, the photographer, the manager, the lawyer, the producer, the copywriter, the accountant, the social media manager, the human resources, the CFO, and so on," wrote one. "Running a business as an influencer is like running any small business—you have to wear so many hats. The difference is that you don't get credit or recognition for any of it—different than small business owners that are seen as hardworking."

An easy solution to this problem? Outsourcing, which many influencers do.

Answering DMs, one of the most time-sucking things influencers do, is also one of the easiest things to outsource. Perhaps unsurprisingly, most influencers, like celebrities, have the people who work for them sign NDAs. So this assistant, like most influencer assistants I have chatted with over the years about their job, has asked me not to reveal her real name or identifying details about her or her influencer. I'll call her Josie.

Josie is proud of her career. She had always been good at social media. When she was nineteen she was hired by a local business to run their accounts and do their influencer marketing. Then, she turned that into a small business consulting for influencers who needed help growing or making brand deals. Eventually, her friend, whose account had blown up, asked her to work for her part-time.

One of Josie's primary jobs is answering DMs for her influencer, for whom she works about twenty-five hours a week. Josie, a stay-at-home mom, does some other things, like making and finding affiliate links, but DMs take up so much of her time that they usually swallow up her working hours. She estimates that this influencer gets approximately fifty to sixty an hour.

"There will always be DMs to answer," she said. "Or something to link. Or something to share. Or investigate. DMs are often like a game of Whac-A-Mole. I answer five, then immediately I refresh Instagram and there's five new DMs plus five new replies to the replies that I just sent. And also people are . . . often dumb. The restraint it takes to not send a hundred 'just google it' replies a day is unreal."

She also is always amazed at how many people reach out to her influencer about their lives.

"It's so crazy to me, but then it shows me that they really feel a connection to my boss because my boss is open and authentic or whatever," she said. "People will write to her with super personal information, and sometimes they're not even looking for advice."

She also has to be careful when people get a little too personal.

"I stay away from anything that's sensitive or could potentially get her in trouble," she said—for example, medical or nutritional advice. "A lot of people write in with personal information, and then my boss can decide whether she wants to respond to it."

Josie can see how much money she is making for her boss. Her boss doesn't have time to go through all of her messages herself, but when Josie does, she will see her boss's inbox get flooded with DMs about, for example, where she got her shoes. She will text her boss, and maybe she will be busy and not respond. So Josie will do the work to find the shoes online and link them out on Instagram Stories. Instantly, she knows, she just made her boss three thousand dollars.

In the beginning, it took a bit for Josie to get the hang of replying as her boss and making sure she sounded authentic to her brand. Like, what kind of punctuation does she use? Does she speak in complete sentences, or does she use a consistent set of abbreviations? Surprisingly, Josie said, the hardest thing to nail is authentic emoji usage. Most people have certain emojis they favor, so assistants like her have to learn which ones their boss typically would or would not use.

Josie has a friend who once was on a team of thirteen people working for an influencer. Her role was to be the main DM responder because if too many of the assistants responded to DMs, it would become inconsistent. Her friend's boss was very intense, she said, and very picky about how the DMs were written.

For example, the influencer wanted to use only the "yellow skin" emojis, instead of picking a skin tone. But after the Black Lives Matter protests in 2020, Josie said, the influencer changed her mind. When Josie's friend used the yellow-skin emoji afterward, her boss was furious. Her boss had decided it was more appropriate for her to use the "white skin" emoji. But Josie's friend still wasn't using the *right* white-skin emoji.

It was an emoji disaster, with consequences. The influencer's assistants were on a demerit system, and Josie's friend's emoji mistakes contributed to her getting too many demerits. She was fired.

Josie doesn't have to worry about that, though. She is close to her boss and considers herself extremely lucky to have her position. The influencer world is kind of glamorous and isn't something she ever expected to be a part of.

"Having a role where I am sometimes, for all intents and purposes, a proxy for this beautiful, perfect-seeming figure is fun and something I don't think I would ever experience personally," she said.

In her community, Josie is one of several stay-at-home moms she is friends with who all do similar jobs for influencers. Sometimes, they can't believe their luck. Josie and her friends take their kids to the park together, watch them play, and answer DMs about beauty products and fancy home furnishings. Josie admits that it sounds like an MLM-scam job when she describes it, but it's totally real. Like Shannon, Josie also got her first connections in the industry through the Utah blogger network, when she lived in the state. Now she has helped several friends get roles similar to hers, just through word of mouth.

"It sometimes blows my mind that I was able to find a job that I could do while waiting to pick my daughter up from school or while I'm in the drive-through line grabbing my kid's happy meals," she said. "Although I feel like I'm constantly working, very rarely do I have to pass up on things in real life because I 'have to work.'"

Josie also thinks that if people understood more about the industry, they would get why people like her boss need assistants like herself.

"These influencers are not just shooting from their hips," she said. "Things are more controlled and calculated than the average follower would anticipate." She added, "Running an account is a full-time operation and the people making huge amounts of money are often truly earning it at the end of the day, in my opinion."

After years of working in the industry and seeing how it's run, Josie has even been considering a bigger career change. For years, it has always been in the back of her mind: Couldn't she do this too?

"It is hard to look at the amount of money influencers can build out of 'nothing' and not try to figure out where you can fit in that space," she said. Especially because she is so intimately involved in the day-to-day.

"I'm so happy for my boss and I love her and she deserves this and everything is so great. But I was literally getting sick to my stomach being like, how? Why am I spending so much time doing this for somebody else?"

Josie has been dipping her toe in the water recently. She has made her own account and is seeing how it goes. Her boss is supportive and even shouted her out once on her page. It may go nowhere, but Josie feels like she has to try.

"I know that the money is there, and, like, I feel stupid for not trying when I know of the opportunity," she said.

For many years, Josie said she felt a lot of pressure to figure out her "brand" and make her own account. But now she has a more relaxed attitude about it.

"If it doesn't work, I'd be fine working in my current job for the foreseeable [future]—I really do enjoy it and most days I feel like I hit the lotto as far as work-from-home slash stay-at-home mom jobs go," she said. "But it's just hard to look at the money that could potentially come in and not try to find that for yourself." Most of her influencer assistant friends feel the same way.

If Josie is a success, she may be able to make more money than she ever dreamed she could. She may even need to hire an assistant down the line. Maybe two. Maybe thirteen. Think about how many people have been hired over the years for influencers, from full teams of DM assistants to Shannon's Poshmark reseller.

Shannon just paid a new worker for something different, though. After two years of attempting to get rid of free stuff she didn't need that was cluttering up her home, she hired someone to just get rid of everything. She paid the person four hundred dollars, and she has never felt so free.

But that doesn't mean Shannon isn't going to still try and get some things comped through her account. The free stuff is really hard to give up, and after all, her five children will all need braces.

"Maybe you can quit when they are in high school," I told her.

"But even then," she said, "it's like, what else can I get?"

10

Shannon turns to me, eyes wide like she's a kid in a candy store. As I trail behind her, she asks me, "Isn't this amazing?" and "Don't you love this place?"

We are in an H&M inside an outlet mall a short drive away from Shannon's house. I'm a little confused by her excitement. Sure, H&M is cool, I guess? But Shannon looks like she's having the time of her life.

We walk from floor to floor, assessing all the clothes carefully. Trailing behind us are her oldest child, Hudson, and her youngest, London. Hudson is a quiet and thoughtful boy, who tenderly keeps an eye on his baby sister without any prodding from Shannon. He also smiles at Shannon's exuberance, although he tells her he doesn't want any new clothes. London joins in, though. The toddler, who just got a new pair of light-up boots from a shoe store and is now stomping around with glee, is happy to play model for her mother. Shannon squeals at an animal beanie in the kids' section, which she plops on London's head. "Isn't this soooo cute?" she asks, before heading back down to the adult section.

Every item of clothing seems to make her more excited than the last. Shannon has been looking for an oversize white button-down shirt to wear as a bathing suit cover-up when she and Dallin go on a planned trip to Miami. She has already purchased a tan bucket hat for the occasion, which she wore as soon as she bought it, and the shirt will be the finishing touch on the look she envisions wearing poolside.

We check out a few shirts before she finally decides on one, but then she's immediately drawn in a different direction. She homes in on a white knit cardigan. "Oh my God!" she exclaims. "Can you believe this is only ten dollars?" In a whirlwind of clothing and glee, Shannon races up to the cash register, chattering away about how she can't believe her good fortune and plucking the hat from London's head, briefly, to pay for it.

As we leave, Shannon explains to me that she gets such a kick out of shopping for herself.

For years, she felt like she couldn't.

Of course, she had the ability and money to buy clothes, but as an influencer, she felt strange doing so. She wanted to please the brands who sent her so much free stuff in hopes she would post it and didn't want to seem ungrateful or do a bad job. Besides, how could she really justify buying clothes she really wanted when she had boxes and boxes of free stuff she could never even wear? So she didn't shop for herself and didn't pick out her own things. After a while, she began to feel like someone else's Barbie doll. Couple that with the fact that she's spent most of the past decade pregnant, and Shannon hasn't felt like her body has been her own for a long time. This simple act of going to H&M and buying what she wants and making herself feel confident is a revelation for her.

She's been thinking lately of retiring. It may be time for her to give it all up. After the 911 incident and fallout, her career has been, in many ways, more trouble than it is worth. She doesn't have the energy to fight with people in her comments or try to convince brands to stick with her after they get bombarded with nasty messages for working with her. She has stopped pursuing new partnerships, running her business as more of a passive entity. If someone reaches out to her, she will do a partnership, but she's no longer hustling as she once did.

"I still take them here and there but . . . it takes me a long time to write back," she said.

Shannon feels freer than she has in years. Her kids are older and are becoming more and more independent. After spending her twenties attached to one baby and then the next, and creating content in her downtime, she feels reborn. She's taking a look at her life and thinking to herself: Who am I? What do I want?

She's been dipping her toe in some new ventures. While she didn't get the gig on *Real Housewives*, it was exciting for her to step outside her usual role of just being a mom. She scored some acting and modeling jobs, appearing in a few commercials, and has even done a little voiceover work. Once, a long time ago before she got married, Shannon had dreamed of being a broadcast journalist, anchoring the local news or something like that. Doing the voiceover work has been so exciting because it makes her feel like maybe she could still achieve that long-ago dream, or at least do something close to it.

Shannon's trying to rebrand. After her identity has been as a mom on the internet for so long, she wants to go back to being just Shannon. Of course, she is and will always be a mom, but she doesn't want that to drive her career any longer. She wants more than just shilling baby toys and car seats. She wants to build a business she loves, and she wants to respect her children's increasing requests for privacy. She has been experimenting with silly Reels of just her own brand of comedy, no kids included. In many of the videos, she shows off her dance moves from when she used to dance in college, occasionally wearing a sports bra and short shorts.

People on her snarker forums have made fun of her for the videos, but Shannon doesn't really care. She's feeling herself, and she gets good comments too. "Maria from the Sound of Music been real quiet since u posted this," wrote one person on a video of Shannon dancing outside among the rolling hills of her backyard, wearing a robe.

If Shannon had her way, she would create a whole new life for herself. She would move out of Utah. She'd move to Florida and build her own business. She dreams of making her own line of swimsuits. And then she would vanish from the internet, deleting her accounts.

"Yeah, that would be my dream," she sighed.

Shannon's generation of mommy bloggers is hitting a crossroads. Many of them are ready to hang it up or are pivoting to new ventures. Dallin sees it happening in their community. He thinks that after more than a decade of living so publicly online, a lot of the OGs in the neighborhood may soon be ready to make a change. After all, they aren't the young moms they once were. Their kids are getting older and busier, and more intent on their own privacy.

A new wave of moms is coming up and taking the mantle. On TikTok, a whole new generation of Mormon Utah influencer moms have popped up, and they share a lot of characteristics with Shannon and her peers when they started. These influencers, like Camille Munday, Miranda McWhorter, and Taylor Paul, are young hot moms with designer clothes, adorable babies, and because it is TikTok, some good dance moves. They even have their own scandals. (Taylor was ostracized from the group of hot Mormon moms known as #momtok after she revealed she and some of her friends were swingers. If you have some time, just google it. Trust me, it's entertaining.)

The fact that they may be ready to move on from full-time content creation doesn't mean that Shannon's generation needs to wither and die for being passé, Dallin said; far from it. The savvy ones will last and evolve. In Dallin's view, the smartest thing a longtime influencer can do is leverage their social media success into a brand, which can slowly become their primary source of income. Several people Shannon and Dallin know have successfully done so. Amber Fillerup Clark, of *Barefoot Blonde*, launched a haircare line, Dae, that is now sold nationwide at Sephora. Rachel Parcell and Emily Jackson have both launched fashion lines, which have been sold at Nordstrom. In fact, almost every big OG blogger has launched some sort of product or collaboration over the past few years, like Caitlin's line with Pink Lily.

"If you really want to have a long-term career, make a lot of money, you've got to get your own product," Dallin said.

It takes the pressure off constantly having to put yourself and your family out there for public consumption, and hustling for brand deals. No one can live like that forever.

It's true, many influencers, especially those who have been in the game for a long time, are burned out. From 2020 to 2022, one of the biggest story lines in the influencer community that I observed was their growing dissatisfaction with Instagram. Influencers had a love-hate relationship with the platform; it enabled their visibility and success but took control out of their hands. It also created a sense of paranoia. It seemed like no matter what influencers did, other people were getting ahead, and doing so, in their minds, nefariously.

One of their biggest complaints was the recent trend of "loop giveaways." These giveaways started with celebrities like the Kardashians and reality show stars and soon trickled down to influencers. If you are unfamiliar with the term, I'm sure you have seen a photo of Scott Disick or Kylie Jenner on your feed, surrounded by something ridiculous like twenty-five Gucci bags or literal piles of cash. The celeb will say something like "Follow these accounts to enter to win all these prizes," with a list of names of smaller accounts that are trying to grow.

These celebrity giveaways are often run by an outside company that recruits the celebrities and participants and handles the logistics. One of the most notable is Social Stance. If you are a smaller account that wants in, you have to pay Social Stance to include you on the list the celebrity promotes (I actually have been recruited to join a celeb giveaway, and Social Stance quoted me five thousand dollars to get in). You pay to get a bunch of new followers. Once you have these new followers, you can get ad deals with bigger brands, or more lucrative deals, because you have a bigger following.

In 2020, influencers and aspiring influencers began to use loop giveaways to grow in earnest. Most of them didn't work with a company like Social Stance, instead simply organizing the giveaways among themselves and pooling their money to buy an enticing prize.

Some influencers on Instagram see nothing wrong with using loop giveaways to grow. But others feel like it is a cheap and unethical way to get ahead. Influencers who had spent years growing their follower account organically were angry to see people who had the cash to enter a new loop giveaway every week skyrocket past them, reaching five hundred thousand or a million followers in a matter of months. These newbies were able to then broker huge brand deals because they had so many followers. To those who had built their platforms "the right way," it felt like people who were cutting corners were being rewarded by the industry, which didn't seem fair.

"When so many influencers cheat (bots, loops, etc) to look bigger and better, those of us who have grown our audience organically get left in the dust," wrote one in my survey. "So trying to keep up with what's happening now and also maintaining a sense of balance in your life is so hard."

Additionally, these giveaways seemed to violate Instagram's terms of service, which state that users should not be "artificially collecting likes, followers, or shares" or "offer[ing] money or giveaways of money in exchange for likes, followers, comments or other engagement." However, Instagram has never really cracked down on loop giveaways in any meaningful way. In May 2020, a spokesperson told me they were investigating after a group of influencers did a giveaway for a car, but nothing ever happened. This apparent flagrant disregard for Instagram's own rules has only made the giveaway haters madder, mainly at Instagram. Influencers felt that Instagram wasn't implementing or enforcing rules that were best for honest creators, and wasn't supporting a healthy business model that helps those who do things "the right way" get ahead.

This, of course, is just one of many complaints influencers have had in recent years about the way they've been treated by Instagram. In the mid-2010s, Instagram had exploded as a content machine, and old-school bloggers like Caitlin were lured over to the platform as a way to grow their brand and audience. By 2021, though, Instagram content had swallowed much of the blogger industry that came before it. Influencers have told me many brands will pay only for Instagram content and look only at Instagram numbers when determining your rate, or they will pay premiums for Instagram. (TikTok is gaining ground slightly. *Morning Brew* reported in October 2022 that several brands were devoting about 25 percent of their influencer marketing budget to TikTok campaigns, with the remaining 75 percent going to Instagram campaigns, whereas they used to devote their entire budget to Instagram.)

Slowly, blogs and other types of content began to fall by the wayside, until many influencers found themselves spending the majority of their workdays on Instagram. Without realizing it, they had given up something crucial. Instead of owning their business on their own platform, they now were subject to the whims of a corporation they couldn't control and that didn't seem to care much for them.

As I discussed in chapter three, for many years, Instagram offered influencers little to no support in building their companies and didn't offer a way to make money directly through the platform (Instagram, of course, would disagree with this characterization, with CEO Adam Mosseri saying in April 2022 that Instagram has "done a lot" to support creators over the past decade). But there were other issues too. Perhaps the biggest is the opacity of what influencers refer to as "the algorithm," or Instagram's tools that determine whose content is shown to whom, and why.

Instagram has tried to downplay how much the algorithm

controls an influencer's success. In a June 2021 blog post, Mosseri wrote that the existence of an almighty algorithm is one of the biggest misconceptions about the platform, saying, "Instagram doesn't have one algorithm that oversees what people do and don't see on the app. We use a variety of algorithms, classifiers, and processes, each with its own purpose. We want to make the most of your time, and we believe that using technology to personalize your experience is the best way to do that."

Mosseri claimed that each individual user is able to tailor their own experience and pick which creators they want to see.

"How you use Instagram heavily influences the things you see and don't see," he wrote. "You help improve the experience simply by interacting with the profiles and posts you enjoy."

I have spoken with several influencers over the years, though, who feel differently. They believe that Instagram does reward certain behavior from creators. For example, influencers tell me over and over that the platform rewards creators who spend more time on the app, whether they are posting stories, responding to messages, or liking other posts. They believe their time on the app directly correlates to how many people see their posts. If they are on Instagram all day, their engagement shoots up. If they take a break, it tanks. Though this is only anecdotal evidence, it has come up so often that creators clearly believe it's true regardless of the company line.

Some of the influencers in my survey pointed to this specifically in discussing the difficulties of their jobs. "There's never really an end or beginning to my days," said one. "Or vacations.

It's all about sharing 24-7 and that can be hard. When you take time off it feels like social platforms can hold that against you, so you constantly feel like you have to keep up momentum or fall behind."

Other factors seem to boost—or tank—engagement. In 2020, Instagram launched its TikTok competitor, Reels, to great fanfare. The company aimed to grow new creators with Reels. As an enticement, it offered "Reels bonuses" of up to ten thousand dollars for influencers who posted them. This was one of the first ways influencers could get paid directly on the app.

Instagram also started heavily promoting Reels in the feeds of its users. I don't need to work in social media to know this, I just had to open my Instagram account. For months, my entire feed was Reels. Reels also became a growth hack, which I observed in my personal Instagram use. I was amazed to watch a girl I knew from my hometown grow from a few thousand followers to more than a hundred thousand in just a few months. All she did was post Reels. Tons and tons of Reels.

Is there anything wrong with taking this growth hack and running with it? No, not really. Instagram was pushing Reels, and if people want to use that to grow their platform, more power to them. But for the old(er)-school, magazine-style influencers who had built Instagram into a juggernaut, it was frustrating. For one, many of them said their audience didn't seem to like this new direction. Followers would complain when they posted Reels. They didn't come to Instagram for TikTok, they wanted the same content they had come to expect on Instagram.

But if influencers didn't play the game, they felt like no one was seeing their posts. Was Instagram even for them anymore?

Caitlin has felt this way for years. Honestly, she never wanted to be an Instagram influencer—that is, *just* an Instagram influencer. She always identified as a blogger, and held on to that even as blog readership slowly declined and Instagram began to crowd out any other type of content creation.

"Instagram is definitely frustrating. I feel like they keep changing things and they're pushing Reels so heavily and people don't want to see Reels," she said. "They want pictures. Engagement and views are so low across the board. I feel like all of my friends I've talked to that are bloggers say Instagram isn't showing their content to anyone. So it is tough and I don't want to put all my eggs in one basket."

Even though Caitlin always wanted to keep her identity as a blogger, for a while, her content was performing well on Instagram. But then she remembers a big change. Around 2016, she suspected that Instagram changed the feed from showing photos chronologically to using an algorithm to determine what posts showed up in followers' feeds. Caitlin's engagement began to plummet. Her photos would "bomb" with scary regularity. The lackluster engagement was a hit to her confidence.

"I just thought it was one of the hardest years because I was pouring everything into my business and Instagram had control over it and the content wasn't being seen," she said. "And I just felt like such a failure. It was like going to work every day and your boss saying, 'You're not good enough, you're not good enough. What you're doing is not good enough.'"

Over time, Caitlin began to realize that it wasn't her who was the problem. The problem, she said, was that Instagram was implementing changes that she felt were benefiting only the company, not the influencers who were sharing content. Once she realized that she was doing everything she could to succeed, she stopped being so hard on herself.

"I used to put so much of my self-worth in my Instagram, like how a certain photo would do," she said. "I think that one really bad year that I had, it just kind of like taught me that, hey, your photo can bomb and it's OK—like, you're still a good person. It doesn't mean that you didn't do a good job on the content. Instagram literally isn't showing people what you're doing."

Caitlin has tried not to focus solely on Instagram as her only means of content creation. She never slowed down on her blog, which she has always found more fulfilling than posting on Instagram. Sure, her blog views are a fraction of what they once were, but she has a core group of readers who reliably keep up with the blog and read and comment on her posts. It's fulfilling to have that feedback and to know it is something she built all for herself. And just because people aren't reading as much doesn't mean that her blog can't be a success. A post she wrote in 2019 about her attempts to conceive her daughter is one of her most successful of all time.

Caitlin has also experimented with TikTok and Reels, which

she enjoys. She likes trying out videos, even if it's not her main form of creating content. Reels have actually been a huge success for her; some of her fashion Reels have the highest conversion rate to sales of any of her content. But she knows her audience doesn't really like them, and she's just playing into what Instagram wants. That's why her audience is seeing her posts, not because she is doing what is best for her brand.

"Those will get high engagement, and I do feel like this helps me," she said. "But if I do post random videos all the time, I think that would make people mad because they don't want to see that. So I try to be strategic about which videos I'm posting."

This sentiment is one I have heard more and more, and it's becoming a rallying cry in the industry. Amber Venz Box of RewardStyle is a huge advocate of not tying your business revenue to Instagram, as is Caitlin's manager, Kirstin.

Kirstin believes that the only way to really have a long-term career in the industry is to own your income streams and diversify them, whether that's through a podcast, a brand, a newsletter, or a blog. Even if they are killing it on every single platform, she tells her clients that they need to look elsewhere and find something they can own outright. It's just too risky to rely so fully on another company.

"You never know what's going to happen with any platforms," she said. "Any single one of them at any given time can have an algorithm change and totally mess up your business."

Kirstin is also a huge advocate of old-school blogging. She's had clients come to her and tell her they plan on giving up their blog, or clients who have never had a blog telling her they don't really see the point of starting one. She always disagrees.

"There's the only place you get to have complete control of your audience," she said. "And that is a precious, precious thing that you should never give up."

Most of the influencers I surveyed are also working to make sure their businesses remain under their control. "I'm acutely aware of the dangers of relying on these social media platforms to make our living, so I'm always thinking about ways we can build businesses that leverage our built-in audience, but can exist independently of our personal brand," said one.

These influencers have varied ambitions and plans. One said her goal was to build up the brand she'd created to the point that she could scale "back on the number of sponsored campaigns I do annually," another planned to try out video content, and a third hoped soon to make her podcast her main source of income. Another said she was simply hoping to "leverage [her] experience and audience as an influencer to create a more solid business."

"I plan to leverage my following into securing bigger investment opportunities that do not rely completely on my social media following," one influencer said. "Not relying on sponsored posts to make a living is my plan. By owning a stake in companies that I post about organically, it will not come across as hashtag-sponsored and I will own a piece of the stake. I am also starting my own brand that I will own and one that cannot be controlled by outside forces."

That's not to say that these influencers all want to quit being influencers. Far from it. They are all extremely proud of what they have built and enjoy what they do. That's why they want to pursue these other projects. They finally recognize their worth and have the confidence to not rely so much on another company for their success.

"As influencers realize the importance of their roles in marketing and advertising, and the impact they have, fees will increase and influencers will move into selling their own products and services more versus only advertising other brands," one predicted.

One put it simply: "I think we're all getting a little tired of being at the mercy of social media algorithms and want to take control of our content and livelihoods."

There's someone who agrees with them: Mosseri. In a 2022 TED Talk, he expressed his view that while the platforms had most of the control over how content was disseminated over the past decade, he believes the next decade will be one in which power will go back to individual creators. Perhaps seeing the writing on the wall, or, in a kinder reading, out of a genuine desire for Instagram to be a place where creators can succeed, Mosseri said he plans to spearhead initiatives on Instagram to help creators monetize their accounts and grow their success. He and Instagram don't see the encouragement of Reels as antithetical to that goal. In their mind, they are just trying to give everyone more opportunities to grow and connect with their audiences in a new way. And audiences seem to like video—after all, just look at TikTok.

According to Mosseri, Instagram is ready to give the power of the internet back to the people who built it.

"What if we imagine a world where creators actually own their relationship with their audience—they didn't rent it, they owned it—and where all of us were invested in their success?" he said. "A world where the platforms acted more like platforms because we can and should do more to support creators."

CAITLIN DOESN'T WANT to be an influencer forever. To be honest, when she said this, it surprised me. After everything she's built over so many years, would she really just give it all up?

Yes, she would. Because while she's proud of herself and she loves her work, being an influencer takes a lot more out of her than a more normal job would. Caitlin can't work just eight hours a day and then focus on her family life. Her life is her work, and in order to be a success, she has to let people into her personal world. After so many years of showing herself fully to the masses online, she is growing tired. She knows it's not sustainable. She can't do it for the rest of her life.

There have been times she has considered quitting before. The first time GOMI went after her. The year when her Instagram success plummeted and she took it personally. She didn't quit then, but she began to realize she may have to someday.

"I don't want to do this forever," Caitlin said she thought in those dark moments. "I don't want my personal life to be pried into forever. And then it got better. But it's always kind of been in the back of my mind, like, do I want to do this forever?"

She's not thinking of nuking her accounts or anything. But she doesn't want to be tied to showing her life to make a living. Caitlin could hop on Instagram from time to time and show little bits and pieces of her life, but not make herself the focus of her livelihood.

"I don't want to always kill myself to be so present on Instagram," she said.

Couldn't she just stop showing the personal side of her life, and keep her accounts strictly fashion or lifestyle? Some may say yes. Caitlin doesn't think so. In her experience, the only way to truly be a successful influencer is to be able to meld the personal and the professional. She knows people who have tried it, and she doesn't find them compelling. Who wants to follow someone when you can't see their soul?

Caitlin has come to the conclusion that there's no way to have the success she does and still keep herself out of it. She knows what the people want. She just doesn't know how much longer she will be able to provide it for them. And if she's not doing her best, if she's doing the job only halfway, she doesn't think she will be able to continue on.

"I'm not happy with something unless I'm putting, like, 100 percent of myself into it, and it would be really hard to only put a little bit of myself and know that it wasn't being successful," she said.

Like many other influencers, Caitlin is thinking of diversifying. Her goal is to create a product. The product could be launched off her success as an influencer, and she could use her platform to get it up and running. Her current idea is a line of hair products that contain only "clean" and natural ingredients but still work as well as traditional hair products.

Once she can launch the brand, Caitlin would like to hire a team to promote it organically and run the brand daily. Then, slowly, she can start to phase herself out. She can stop posting on Instagram Stories as much. She can share less and less of herself. She may pop onto Instagram or the blog from time to time, especially in support of her brand, but she won't be reliant on sharing herself to make an income anymore. Her brand and her business will no longer be her, it will be her company. That is the goal.

Her long-term dreams are rather simple. She can picture the life she wants to lead in five years. She and Chris will have three kids. She will be able to build her schedule around them, dropping them off and picking them up from school, and shuttling them to any activities they have. In between, she could blog if she felt like it, or do anything she needed to do with her product line. Hopefully, it would be rather self-sufficient. When her kids are home, though, she wants to be present with them. She wants to be a mom, and she wants to focus on them, not herself.

"So like, a really good balance of work and time with my kids, but also really a successful brand that's kind of running itself that I just have to do a few things for, but not necessarily, like, pouring my soul into, like, every second," she said.

Until then, Caitlin is thankful for the life she has built. It's not perfect and it has its own challenges, but it is all her own.

"There's a lot of negative things about influencing, but my job allows me to be with my daughter all day and I still work and allows me to travel when I want to. And being my own boss, I can take a vacation day if I wanted to," she said. "It's just it's allowed me to live a really cool life and I'm so thankful for it. And, like, I want to recognize at the end of the day, like, it's the people that follow me that have allowed this to happen and I'm so thankful for them and I never want to forget that it's because of them."

MIRNA IS MAKING more money now than she ever expected she could, and it's the type of money that's life-changing—for Mirna and her entire family. She thinks for the first time she may be able to own property. She's just started looking at lots of land, like the parcel I accompanied her to look at, because of her big dream.

She explains it to me as we drive through the Vermont countryside, her voice melodic as she shares her vision. At the end of the day, all of her work is leading up to this. Her higher purpose, her calling, which could be made possible by this strange yet wonderful new path she has started on. Mirna's influencer career isn't slowing down—in 2021, she made more money than she ever had before, and got even more opportunities, with bigger and bigger brands.

First, Mirna will build a house. It will be big enough to fit her whole family, so they can all stay comfortably with her. She wants to host big gatherings, holidays, and parties, where everyone can stay with her and not pay for a hotel. A three-bed, two-bath would probably work. She wants the house to be completely sustainable, or as sustainable as possible.

The house is just one aspect of her plan, though. The first thing she would actually do when she bought the land is build trails, for running, walking, and biking. She wants enough acreage to be able to have a loop that could at least be a mile long, and would be comfortable for anyone to run on. She could start holding races on her property—5Ks and 10Ks, for both walkers and runners. Maybe, she could even do a 12- or 24-hour trail run, in which people complete an ultra-running distance by circling the trails as many times as they choose.

Mirna wants to host these events on her property because she wants to do them her way. She wants them to feel completely inclusive, where anyone of any race or size could come and participate and feel as happy as she does when she is outside. It would be an event where people can run and have fun and not be judged.

To know Mirna, though, is to know that she is never thinking merely one or two steps ahead. She is always pushing for the next thing, the bigger goal. She has grand ambitions, and as we drive, she reveals her ultimate dream to me.

As a kid growing up in Brooklyn, Mirna fell in love with the outdoors by going on just a few camping trips. Those experiences changed the direction of her life, sparking a love of the outdoors that could never be quenched and that led to her current career. Everything she has built for herself exists because she had the opportunity, however small, to find a passion. She wants other people to have the opportunity to find that as well.

So eventually, she will build it. She will find a plot of land big enough, or she will buy a second one. On it, she will build the ultimate outdoor retreat for people to come and relax, and get acquainted with the outdoors and outdoor sports. She will have a ton of different ones to try—not just running but also biking, maybe even skiing. She wants the retreat to specifically be geared toward people of color, so they could try out sports without the fear of judgment and pressure that comes with being in a predominantly white space. And she would like to figure out a way for the sports and activities to be free so that people could come who otherwise would not be able to afford them.

For Mirna, becoming an influencer has made dreams like these a reality. Starting this new career, and taking this leap, has helped her live her best life every single day. Now she wants to help others have the same experience. Even she can't believe it all started with a blog.

"Running has brought me this platform, and the ability to live my [running] life out loud, in my big black female body," she wrote once on Instagram. "I celebrate the opportunity to leave a mark, a legacy, or even to make a small ripple—but of what? Of acknowledgment, acceptance, of how when you extend yourself to others, you create that space in them to do the same, of how when you make the effort to set aside any preconceived notions about a person's identity, and make no mistake, I'm talking about race, you might learn a bit more about yourself, them, and well, the world around you."

11

Miranda McWhorter never expected to get famous on the internet.

It happened so organically. One minute, she was a normal Utah wife and mom. Like so many of her peers, she had married when she was fresh out of her teens and had her first child, son Brooks, at twenty-two. She didn't follow any influencers super closely, although she liked a couple well enough. And she had no interest in getting this app called TikTok.

"I thought it was just another dumb app to waste your time on, which, I wasn't entirely wrong," she told me in the fall of 2022.

Then, in March 2020, she finally got bored enough to download the app. All of her friends were obsessed with it and were posting more and more videos of themselves. Miranda decided to join in. Her first video, captioned "hey, we're new here," showed Miranda and Brooks changing from sweats to all dressed up and ready for the day. She hashtagged the video #mom.

From then on, Miranda started to post about her life. She shared her daily life with Brooks, her makeup tips, and her outfits, tagging brands to show off what she was wearing. She and her friends Camille Munday and Taylor Paul filmed videos dancing to the latest TikTok song trends, hashtagging the videos #MomTok (you may recognize these names from the previous chapter, or the infamous swinger scandal that tore the group apart in May 2022). Her first video to hit more than a million views featured idyllic scenes from her relationship with her husband, Chase, their engagement, wedding, and gender reveal, set to music. "So in love," she wrote.

Pretty soon, her new followers had questions. Where were her pants from? Where'd she get that flannel? How did she curl her hair? So Miranda started doing tutorials, posting fashion recommendations, and doing hauls from brands like Shein. Then, brands started to reach out to her. When she secured her first brand deal, she realized that her account could actually be a new career.

"I never intended for this to become my job but when my account started growing and I began to make money, that's when I realized it could really be a full-on career," she said.

Ah, the circle of life. Or in this case, the circle of beautiful, young, Mormon influencer moms. Because it turns out, Mormon mommy bloggers aren't dying off or becoming irrelevant. They are evolving, and a new generation has stepped onto the stage.

Miranda now has more than a million followers on TikTok,

her main platform, where she presents traditional lifestyle and parenting content like Caitlin and Shannon do, but in a mostly video format. Riding in on the success of the women before her, she has launched a successful content creation career in a very short period of time and is inspiring millions of young women to be just like her. She is part of the next generation of influencers—they're a little savvier and a little more versatile, and are having a lot of success.

Miranda knows that social media can be both, as she says, a blessing and a curse. But like the generation of motherhood influencers before her, she is thankful for the symbiotic relationship her life has with her career. She not only is building a platform for herself to support her family, but she also is finding community and kinship online. She feels lucky to be a mom in a time when she can find so many resources, something she thinks previous generations would have killed for.

"I think having a community on social media that can help you help your little ones through that process is incredible," she said.

Many of the difficulties of her career also feel familiar. Miranda has at times struggled with balancing her life as a mother with her job in content creation. It's not the hardest job in the world, she concedes, but it's not super easy either. The constant hate and criticism she gets as a public figure is the hardest part, by far.

"The mental toll this career takes on you makes it feel very taxing at times," she said.

Since I started reading GOMI back in 2011, people have been predicting that the end of influencing as a career path is right around the corner. A lot of people, and media publications, have written about a proverbial "influencer bubble" that is bound to burst one of these days.

These articles, I've found, rarely have any hard evidence to back up these claims that a bubble exists. The main assumption they make is that eventually consumers will move on. The argument goes: Most influencers buy followers anyway and their careers are therefore illegitimate, so it's only a matter of time before followers catch on and stop engaging with them. Besides, no one believes influencers actually like the products they are selling. Eventually, the money will dry up too.

"Consumers can see if someone honestly cares about a product or whether they are just trying to push it," Anders Ankarlid, the chief executive of an online stationery retailer, A Good Company, said in a 2019 *Wall Street Journal* article. "The bubble is starting to burst."

This prediction hasn't really panned out, at least not yet. After all, when *The Wall Street Journal* wrote that article, we were just starting to see the emergence of TikTok as a major player in the creator space; in 2021, the most-followed influencer on TikTok, Charli D'Amelio, made $17.5 million. The industry has grown steadily year over year, and by 2025, according to one report, the influencer marketing market is expected to reach $24 billion in revenue.

Still, people are predicting, or perhaps hoping, the industry will cease to exist. This was compounded by the explosion of TikTok in the years I wrote this book, which many people seemed to believe meant that the only creators who mattered anymore were the ones who went viral there. The chatter even got to me at one point. "Am I writing a book on a topic that in five years is going to be completely irrelevant?" I wondered to myself.

Then, I watch someone like Miranda, and I realize I have got it all wrong. Magazine-style influencers aren't dying, they are evolving. Miranda posts on TikTok and Instagram, depending on if she wants to create in a video or written format. Some of her peers are launching newsletters, blogs, and podcasts to accompany their TikTok accounts. And the next generation is building even more sustainable—and perhaps more lucrative—businesses for themselves than their predecessors.

In fact, influencers are impacting and changing a whole swath of industries. The ability to sell your brand to push products online is a bona fide business skill set at this point that almost every industry utilizes. Think about how many businesses are following their lead. Every time you see some brand making goofy TikTok videos to try and sell products, they are just copying influencers. When C-list celebrities or reality stars from the '00s get tired of Hollywood, they can rebrand themselves as content creators online, sharing recipes or just their new lives as parents.

Even journalists like me have begun marketing ourselves on social media platforms in the style of influencers, disseminating our work primarily through social media.

Shannon has gotten to watch the next generation rise up firsthand. A girl named Indy Severe used to watch her kids when they were young, and she and Shannon have stayed in touch. After graduating from high school, Indy realized she had a talent for videography, and took off traveling all over the world. Through her breathtaking videos of her travels, Indy grew her platform as an influencer. When she and her boyfriend found out they were expecting a baby, she moved back to Utah for good.

Now Indy, who turned twenty-five in 2022, is a lifestyle influencer, and part of her persona is her life as a young hip mother to her son, Seven. Indy is a different type of mommy blogger than Shannon; she is a little edgier and has the deadpan meme speak that is native to Gen Z. But she is one nonetheless. Like many Gen Z creators, she has diversified quickly. She owns a clothing line, Lonely Ghost, which even has a brick-and-mortar store, Ghost Grocery, in Provo.

Indy also blogs. She writes often on her website, long, almost stream-of-consciousness-style missives that are poignant and heartfelt. In one, a letter to her son called "Dear Seven," she pours out her fears and hopes for her baby and discusses the highs and lows of her journey with motherhood thus far. The post reminds me of the blog posts I used to read in 2010, before influencing grew up.

This new crop of influencers also has an advantage that the

ones before didn't have. They didn't have to create an entire industry from scratch. Thus, they can improve upon it. Many lifestyle influencers nowadays try to be transparent with their audiences, and are open about the work it takes to run their businesses.

On TikTok, one of the most fun trends for followers to look through is "#PRHaul," in which Gen Z influencers share how much stuff they get for free from brands. Lauren Wolfe, living in New York, frequently posts such videos. In one, in which she shares her favorite packages of the week, she makes the video look like all the packages are dropping in from the sky.

"I'm crying!" she says when she shows off a box of roses. "I don't even drink, but my friends love me," she says, unboxing a huge PR package from an alcohol brand.

The influencers are making these videos, well, because it's good content. It's a big shift from the early days, when bloggers felt like they had to close off the business side of their accounts from follower scrutiny.

"I always used to be so curious about what influencers would get for free," Gen Z lifestyle influencer Kate Bartlett said in one #prhaul video she posted to TikTok. "And now that I actually get PR, it genuinely shocks me sometimes what I get in a week."

In general, young influencers and their audiences alike tend to be savvier about the internet in general. While millennials had to muddle through the weird new world of living life out loud on the internet, it is second nature to the creators taking over. They don't really care that their baby photos are online, and they share details about their childhood traumas in joke videos on TikTok. For Gen Z, there is no "culture" and "internet culture." They are one.

That also means that when many look for careers, they may find influencing. Why would a young girl with dreams of being a fashion editor not want to follow Caitlin's path? Isn't it more freeing to run your own business, to do your own thing, than be beholden to the corporate structure of a fashion magazine? These young women know that they have options. They don't have to rely on others to make their careers for them. They can make their own.

The influencers I surveyed didn't think the industry was dying. In fact, they all see more opportunities than ever. Brands are getting savvier about working with more niche creators, and there is space for anyone. One influencer told me that if she could give advice to an inspiring creator, she would encourage them to just try.

"This industry is still in its infancy and there's plenty of space for tens of thousands more, if not hundreds," she said. "Stop worrying so much about competition and focus on being fully you—because that is the only way of being unique. If you're not clear on who you are and what you're passionate about, you won't find your space online."

"Be completely yourself, don't seek approval from strangers, and have a backup plan—remember that people are watching and learning from you; you have a real and solid influence, so use it for good, not just to receive a paycheck," advised another.

Even though she's not sure how much longer she will want to

be a part of it, Caitlin also thinks that the industry she helped to build will last for many generations to come.

"I think it will continue forever. I really do," she said. "I think it might look different ten years from now. There might be a different platform that is really popular . . . but I don't think influencers are going anywhere."

The influencer industry is not perfect. We have a long way to go before influencers of all backgrounds are paid fairly. People still act like their lives are perfect, and constant social media usage can still negatively impact mental health and body image. Kids are still being shown, and even exploited, on their parents' accounts without any rights or protections. And then there is the can of worms of the problems with Facebook, Twitter, and Instagram that could fill a whole other book.

However, now that this industry has been established, maybe the next generation can improve upon it. Some are already starting to try and make things better. In February 2022, Chris McCarty, a high school student from Seattle, launched a campaign aimed at trying to get more protections for children on the internet. McCarty, whose pronouns are they/them, got interested in the issue after seeing the Myka Stauffer story, *GeekWire* reported, and decided to make the issue the centerpiece of their project for their Girl Scouts Gold Award.

The story reads: "At McCarty's prompting, Washington lawmakers recently proposed House Bill 2032, which states: 'Some children are filmed, with highly personal details of their lives shared on the internet for compensation, from birth. In addition to severe loss of privacy, these children receive no consideration for the use and exchange of their personal property rights.'

"Rep. Emily Wicks, a Democratic lawmaker from Everett, sponsored the legislation. 'You start to think about how that [vlog] is affecting a child's life and their future,' she said."

It's not shocking to me that the possible solution to the question of how to handle children on the internet would come from a member of the next generation. What about the generation after them? These are the children who will have actually gone through the experience of having their highs, their lows, their first crawl to their first days of school to their proms documented on the internet. Many of these children are entering their teenage years now, and soon they will be able to tell their own stories. What will they say?

It may be too soon to tell. But all this movement on the matter is encouraging. The fact that state lawmakers are actually discussing and seriously thinking about the issue means that society is beginning to view content creation as an actual business that can have both positive and negative consequences. After all, in order to fully regulate something, you must first understand and respect it as a phenomenon worthy of serious attention.

When I first decided to write this book, I talked to agents and editors in the book-publishing world, who all asked me one question: What do you want to say with this book? What's your angle here?

I kept returning to one thing: Influencers matter. I repeated it over and over again, in meetings and pitches. Influencers matter. Influencers matter. I want to write a book because I believe that influencers matter.

Of course, everyone told me I couldn't base an entire book on such a simple concept, and I fleshed it out over time. As I sit here now, though, deciding what my final words on this subject should be, I keep coming back to those two words. It's simple but it deserves to be said.

That's because all I have been hearing for the past decade is how influencers don't really matter. They are silly, frivolous, soon to be outdated, and washed up. They aren't having any real impact on the world. And on and on. The industry is overlooked and discounted.

I wrote this book to show you why conventional wisdom isn't true. Influencers do matter, and they are having an impact, whether positive or negative, on all of us who consume what they put out into the world. It can be as simple as buying a new pair of shoes for a first date, or as groundbreaking as introducing the concept of white privilege, but every day millions of women in this country are learning and growing and consuming things from influencers. It's a profound shift. An industry that did not even exist fifteen years ago now drives our spending habits, our fashion choices, our workouts, our recipes, our humor, and our news consumption.

When I think about influencers and their impact, I think about Mirna. I think about how growing up, I never saw anyone who looked like her in an ad for a sportswear company. Now Mirna is on a Lululemon billboard in New York City. When my daughter walks around our city, she will look up and see Mirna as a representation of a female athlete. And women like Mirna are getting a seat at the table and can use her as a voice to get their opinions heard—to finally express what it's like to not be seen, and how companies can be more diverse in every form imaginable. That matters.

And I think about Shannon. I think about how motherhood is messy and complicated, and how for many years no one talked about that. I think about how there are moms out there who relate to Shannon's feelings about motherhood, who get that sometimes it's hard to make sure every child looks perfect, but they are good moms anyway. I think about the tribe of moms sitting together in the comments sections of Shannon's blog posts, laughing about her adventures as they rock their babies to sleep in shirts covered with spit-up. A mom who got an idea for a new product from something Shannon recommended? Or who just felt less alone by her sharing her life? That matters.

And I think about Caitlin. I think about how sometimes life can be overwhelming, and it's nice just to have a friend "in your pocket" who gives you recommendations for the best hair products, the cutest accessories, or the easiest dress. I think about how sometimes I get legitimately stressed out about how to put an outfit together or what to wear, and reading blogs like Caitlin's helps me. Or I think about how sometimes, a new purse or lipstick can brighten someone's day. And how knowing someone like Caitlin, who seems to have it all and struggles too, can help people feel less alone. That matters.

When I asked Caitlin during one of our final calls what she hoped her followers had taken away from all her years of blogging, she immediately teared up. All she hopes is that she has mattered to them, in a good way.

"I hope that it's not something they learned like a makeup hack or something, I hope it's something deeper than that. I just . . . I hope that they see me as a friend and a positive influence in their lives. And that I have some sort of positive impact on them."

That's influence. And that matters.

ACKNOWLEDGMENTS

Before I begin, I'd like to express my sincerest thanks to you, my reader. Over the past decade of my writing career, I have been repeatedly humbled by the amount of support and enthusiasm I have received from people from all walks of life. Every comment, Instagram DM, email, and tweet I get from readers excited about my work means so much to me, and I hope you realize how profoundly it moves me. Thank you so much.

This book has truly been a team effort from so many. First and foremost, I want to thank my interview subjects. Caitlin Covington, thank you for letting me into your heart and mind, your candor, and your willingness to always engage with any question I asked of you. Mirna Valerio, thank you for the best introduction to Vermont I could have ever experienced, your honesty and openness, and for sharing all your wisdom with me. Shannon Bird, thank you for the most fun "tour" any blogger fan could want, your endless enthusiasm about this project, and your willingness to always share what's on your heart. I recognize how scary it can be and how vulnerable it can feel to open up to a journalist like me, and I hope you three realize how thankful I am. I could have never done this project without you. Special thanks also to Ali Pimental, Margaux Nissen Gray, and Kimberly Eagle for coordinating many, many interviews, and to Kirstin Enlow for not only serving as a coordinator but also agreeing to be an interview subject. To my other interview subjects: Dallin Bird, Ayana Lage, Rossana Burgos, Miranda McWhorter, Amber Venz Box, Matthew Kirschner, Michael Heller, and those who shared with me anonymously. Thank you for your invaluable insight.

This book literally would not exist without my editor, Merry Sun, who sent me an email in 2019 and asked if I'd be interested in writing a book about influencers and whose adept hand, smart edits, and clear-eyed view of how to tell this story has shaped this book from start to finish. Merry, I can't thank you enough for believing in me and reaching out to me, and for every idea, big and small, you have graciously given me. I'm so glad you sent me that email. It also would not exist without my incredible agent, Leila Campoli, who took my extremely nebulous ideas about what I wanted this book to look like and deftly guided me into creating a vision that was able to help me see clearly what this work could be. Leila, your guidance and support in every aspect of my career since we met has been invaluable to me, and I feel so very lucky to have you in my corner. Both of your belief in me in my work has given me so much confidence and strength, and I can't thank you enough.

I have been so lucky to work with an amazing team at Portfolio, who have believed in this book from day one and have blown me away with their support and enthusiasm for this project. Thank you to Veronica Velasco for your smart commentary and guidance through all the logistics of creating a book. Thank you to my incredible publicists, Stefanie Rosenblum Brody and Lauren Monahan, and marketing team, Jacquelyn Galindo and Mary Kate Rogers, for having so much love for this book and helping me get past the nervousness of actually releasing it in the world and turning that into excitement. You all have made this process such a joy and so much fun. To Sarah Brody for designing the perfect cover with a dream font, and Mike Brown for expertly managing the process of turning this into a finished book. Thank you to Laurie Flynn for your expert fact-checking and the peace of mind it has given me. Thank you to Alice Lawson and Olivia Handrahan from Gersh for supporting me on this project and others, and for always having my back. I also want to extend a special and heartfelt thanks to those who graciously read early versions of this manuscript and provided their endorsements: Nora McInerny, Jo Piazza, Grace Atwood, Kaitlyn Tiffany, Vanessa Grigoriadis, and Doree Shafrir. I am blown away by your generosity of time and enthusiasm to support this project.

I will forever be grateful to *BuzzFeed News*, where I have been able to experiment, grow, and evolve in so many ways as a journalist and a person. This book would never have existed if I never worked there. Special thanks to the people who have made *BuzzFeed News* such a great place for me to work and who have taught me so much over the years: Lisa Tozzi, Mary Ann Georgantopoulos, Tanya Chen, Lauren Strapagiel, and my entire social news team, Jessica Simeone, Karolina Waclawiak, Tomi Obaro, Estelle Tang, and the culture team past and present, Jason Wells, Ben Smith, and Mark Schoofs. My love for journalism was born when I was a student at the Annenberg School at the University of Southern California, and I'd like to extend a special thank-you to all of the staff members there who taught me so much.

As important as everyone in my professional life has been in the development of my career and this book, those who have supported me in my personal life have had just as big an impact. I consider myself truly fortunate to have found many treasured friends who are constant pillars of support for me and give me the confidence to try new things and put myself out there. Also, they make my life fun and make me happy every day. Christy, Tessa, Broome Street girls (Dani, Miche, Alex, hope you liked your shout-out!), the best former and current NYC friend group chat (Allison, Kaylan, Kate), and many more, thank you for being the best friends a girl could ask for.

Thank you to my extended family (McNeals, Mansfields, Wylies) and the families I married into (Noonans and Henrys) for the constant love and support. I feel lucky to have so many cheerleaders in my corner. To my grandparents for being my biggest fans, and special thanks to Grandma for always encouraging my love of reading and taking me to the bookstore. I love you endlessly.

To my siblings, Bobby, Addie, Jon, and Em, how lucky am I to have you as not just my family but my best friends. You make me laugh harder than anyone and your support gives me a strength and stability that has shaped my entire adult life. I don't know where I'd be without your constant presence. To my beloved nephews JD and Mac, you bring me so much joy and make me smile every single day. I love you all so much.

To my parents, who have given me everything. From Mom illustrating my first-grade short stories for "publication" to Dad printing out and binding the "novel" I wrote in middle school, you have supported my love for reading and writing from the very first time I said I wanted to be an author. I don't know if I really had the confidence to believe it would ever happen, but you always believed I'd write a book. Turns out, you were right. There's no way it would have ever happened without you. For this and so many other reasons, this book is for you.

To Violet, most of this book was written while I was pregnant with you, and I hope that someday you think that's something to be proud of. No matter what I accomplish in my professional life, it doesn't hold a candle to my days sitting and laughing with you. I've never been happier than I am being your mom, and I love you so much. And to Buffy, even though it's silly to thank a cat in your book, I want it in print how much I love you.

Finally, to Brian, for everything. In every stage of this process, from the initial idea to as I write this today, you are my first text or call, my favorite person to have a glass of champagne with, and the only person who can truly ease my imposter syndrome. Your love and steadfast belief in me has transformed me from someone who constantly doubted her own ambition and talent to someone with the confidence to take big swings and lean in to my dreams. How could I do any of this without you? It would be impossible. I love you so much, thank you. YMBB forever.

NOTES

Chapter 1

10 **others in the media:** Lorenz, Taylor. "The Real Difference between Creators and Influencers." *The Atlantic*, May 31, 2019. https://www.theatlantic.com/technology/archive/2019/05/how-creators-became-influencers/590725/.

12 **where influencers reign:** Bain, Phoebe. "For Influencer Marketing, Instagram Isn't the Only Show in Town Anymore." *Marketing Brew*, October 11, 2022. https://www.marketingbrew.com/stories/2022/10/11/for-influencer-marketing-instagram-isn-t-the-only-show-in-town-anymore.

13 **on Instagram were women:** Statista Research Department. "Distribution of Influencers Creating Sponsored Posts on Instagram Worldwide in 2019, by Gender." Statista, August 13, 2021. https://www.statista.com/statistics/893749/share-influencers-creating-sponsored-posts-by-gender/.

15 **32 percent of teenage girls:** Wells, Georgia, Jeff Horwitz, and Deepa Seetharaman. "Facebook Knows Instagram Is Toxic for Teen Girls, Company Documents Show." *Wall Street Journal*, September 14, 2021. https://www.wsj.com/articles/facebook-knows-instagram-is-toxic-for-teen-girls-company-documents-show-11631620739.

17 **female-owned and -consumed:** Statista Research Department, "Distribution of Influencers Creating Sponsored Posts."

Chapter 2

26 **The sub head of the article:** Brant, John. "Ultra: Mirna Valerio." *Runner's World*, June 20, 2018. https://www.runnersworld.com/runners-stories/a21070665/ultra/.

28 **The resulting article:** Bachman, Rachel. "Weight Loss or Not, Exercise Yields Benefits." *Wall Street Journal*, February 9, 2015. https://www.wsj.com/articles/weight-loss-or-not-exercise-yields-benefits-1423505611.

34 **"Hello blog-world":** Covington, Caitlin. "Hello World!" *Southern Curls & Pearls* (blog), April 2011. https://www.southerncurlsandpearls.com/hello-world/#/.

38 **The hunt took its toll:** Covington, Caitlin. "I've Got an . . . Announcement." *Southern Curls & Pearls* (blog), September 2012. https://www.southerncurlsandpearls.com/announcemen-2/#/.

Chapter 3

44 **And it has created:** Geyser, Werner. "The State of Influencer Marketing 2022: Benchmark Report." *Influencer Marketing Hub* (blog), March 2, 2022. https://influencermarketinghub.com/influencer-marketing-benchmark-report/.

47 **The Talent Resources website:** "Our Team." Talent Resources. Accessed October 19, 2022. https://yehuda-neuman.squarespace.com/our-team#:~:text=Founder%20and%20CEO&text=Michael%20was%20the%20Manager%20for,talent%20and%20founded%20Talent%20Resources.

50 **"I'm writing to you from my new apartment":** Covington, Caitlin. "New Apartment!" *Southern Curls & Pearls* (blog), November 2012. https://www.southerncurlsandpearls.com/new-apartmen/.

55 **"I thought my world had ended":** Covington, Caitlin. "Life Updates." *Southern Curls & Pearls* (blog), April 2014. https://www.southerncurlsandpearls.com/life-updates/.

57 **As an undergrad:** Halperin, Shirley. "How Rachel Zoe Became Hollywood's Most Powerful Fashion Player." *Hollywood Reporter*, March 9, 2011. https://www.hollywoodreporter.com/news/general-news/how-rachel-zoe-became-hollywoods-165967/.

62 **Held over two days in Dallas:** Mari, Francesca. "The Click Clique." *Texas Monthly*, September 2014. https://www.texasmonthly.com/articles/the-click-clique/.

Chapter 4

69 **When Shannon first met Dallin at a pool party:** Bird, Shannon. "We Meet," *Bird a La Mode* (blog), n.d., http://birdalamode.com/we-meet/.

71 **"I want all my readers":** Bird, Shannon. "The Birds of a La Mode," July 29, 2016. https://webcache.googleusercontent.com/search?q=cache:rAAp-WfQJxEJ:https://birdalamode.com/page/102/&cd=1&hl=en&ct=clnk&gl=us.

78 **In 2021, I wrote a story:** McNeal, Stephanie. "How These Instagram Baby Clothing Brands Became Collector's Items." *BuzzFeed News*, October 28, 2021. https://www.buzzfeednews.com/article/stephaniemcneal/instagram-baby-clothes-kyte-baby-posh-peanut.

82 **"Ok, so based on comments"**: californianative. "Birdalamode." *Get off My Internets* (blog), July 15, 2013. https://gomiblog.com/forums/beauty-fashion-bloggers/bird-ala-mode-shannon/.

83 **"Bird a la Mode is one"**: Anonymous. "Birdalamode." *Get off My Internets* (blog), July 16, 2013. https://gomiblog.com/forums/beauty-fashion-bloggers/bird-ala-mode-shannon/.

84 **GOMI was started**: Van Syckle, Katie. "'It Put Me on Antidepressants': Welcome to GOMI, the Cruel Site for Female Snark." *The Guardian*, January 21, 2016. https://www.theguardian.com/lifeandstyle/2016/jan/21/gomi-blog-internet-comments-women.

86 **In a 2012 interview**: Orsini, Lauren Rae. "Get off Her Internets: Blogger Alice Wright Bites Back." *Daily Dot*, August 14, 2012. https://www.dailydot.com/unclick/get-off-my-internets-alice-wright-interview/.

90 **To demonstrate, Shannon recalled**: Wright, Alice. "It's Safety First at the Bird Family Christmas." *Get off My Internets* (blog), December 6, 2021. https://gomiblog.com/its-safety-first-at-the-bird-family-christmas/.

92 **Sarah McRae**: McRae, Sarah. "'Get off My Internets': How Anti-Fans Deconstruct Lifestyle Bloggers' Authenticity Work." *Persona Studies* 3, no. 1 (2017): 13–27. https://doi.org/10.21153/ps2017vol3no1art640.

96 **"This is why I"**: Anonymous. "Southern Curls and Pearls / Caitlin Covington / cmcoving." *Get off My Internets* (blog), September 22, 2014. https://gomiblog.com/forums/beauty-fashion-bloggers/southern-curls-and-pearls/page-6/.

96 **They focused on a photo shoot**: Covington, Caitlin. "Our Christmas Photos." *Southern Curls & Pearls* (blog), December 2014. https://www.southerncurlsandpearls.com/couples-christmas-card-photos/#/.

96 **"i find it really inauthentic"**: Anonymous. "Southern Curls and Pearls / Caitlin Covington / cmcoving." *Get off My Internets* (blog), December 28, 2014. https://gomiblog.com/forums/beauty-fashion-bloggers/southern-curls-and-pearls/page-6/."

97 **"how strong of a person I am"**: Covington, Caitlin. "Southern Curls & Pearls." *Southern Curls & Pearls* (blog). Accessed December 6, 2022. https://www.southerncurlsandpearls.com/a-look-back-at-2013/#/.

97 **"Probably a bad sign"**: HLBlech. "Southern Curls and Pearls / Caitlin Covington / cmcoving." *Get off My Internets* (blog), December 29, 2014. https://gomiblog.com/forums/beauty-fashion-bloggers/southern-curls-and-pearls/page-7/.

Chapter 5

105 **After all, experts**: Geyser, Werner. "The State of Influencer Marketing 2022: Benchmark Report." *Influencer Marketing Hub* (blog), March 2, 2022. https://influencermarketinghub.com/influencer-marketing-benchmark-report/.

106 **In 2011, YouTube**: Pickett, Tom. "Supercharging the 'Next' Phase in YouTube Partner Development." *YouTube Blog*, March 7, 2011. https://blog.youtube/news-and-events/supercharging-next-phase-in-youtube/.

107 **Bloomberg reported in 2018**: McCormick, Emily. "Instagram Is Estimated to Be Worth More than $100 Billion." *Bloomberg*, June 25, 2018. https://www.bloomberg.com/news/articles/2018-06-25/value-of-facebook-s-instagram-estimated-to-top-100-billion#xj4y7vzkg?leadSource=uverify%20wall.

108 **and initiatives**: "New Ways for Creators to Make a Living." *Instagram Blog*, June 8, 2021. https://about.instagram.com/blog/announcements/creator-week-2021-new-ways-for-creators-to-make-a-living.

108 **They committed $1 billion**: "Investing $1 Billion in the Creator Community." *Facebook Blog*, July 14, 2021. https://www.facebook.com/creators/investing-one-billion-in-the-creator-community.

110 **"Mountain biking continues to be scary to me"**: Valerio, Mirna (themirnavator). "First of all I'm fine. . . ." Instagram post, October 17, 2021. https://www.instagram.com/p/CVI87IkvL6d/?hl=en.

111 **company Influencer Marketing Hub predicted**: Geyser, "The State of Influencer Marketing 2022."

111 **According to Neal**: Schaffer, Neal. "The Top 25 Influencer Marketing Statistics You Need to Know in 2022." *NealSchaffer.com* (blog), October 8, 2022. https://nealschaffer.com/influencer-marketing-statistics/.

112 **"a majority of consumers"**: Schaffer. "The Top 25 Influencer Marketing Statistics You Need to Know in 2022."

112 **spokesperson for Puma**: Duboff, Josh. "See the First Ad from Kylie Jenner's Puma Campaign." *Vanity Fair*, March 14, 2016. https://www.vanityfair.com/style/2016/03/kylie-jenner-puma-campaign-first-image.

112 **for hawking Chanel**: Moss, Hilary. "Chanel Paid Brad Pitt a Reported $7 Million for His 'Services.'" *The Cut*, October 5, 2012. https://www.thecut.com/2012/10/brad-pitt-chanel-no-5.html.

112 **for advertising McDonald's**: Silverman, Stephen M. "Timberlake: McDonald's $6-Million Man." *People*, September 8, 2003. https://people.com/celebrity/timberlake-mcdonalds-6-million-man/.

112 **commercial for Depend**: Heger, Jen. "Depend on It! Lisa Rinna Banked $2M for Diaper Commercial." *Radar*, April 3, 2015. https://radaronline.com/exclusives/2015/04/lisa-rinna-depends-commercial-salary/.

Chapter 6

131 **wrote KSL in their story**: Egan, Ladd. "Utah Mom Calls 911 for Help with Baby Formula." KSL TV, February 13, 2020. https://ksltv.com/430913/utah-mom-calls-911-for-help-with-baby-formula/.

131 **featured on CNN:** Lee, Alicia. "A Mom Called 911 in the Middle of the Night Because She Desperately Needed Baby Formula. The Police Delivered." CNN, February 18, 2020. https://www.cnn.com/2020/02/18/us/police-respond-911-baby-formula-trnd/index.html.

140 **wrote for Mashable:** Silva, Christianna. "Influencers Aren't Going Anywhere. So What Does That Mean for Today's Teens?" Mashable, October 27, 2021. https://sea.mashable.com/life/18021/influencers-arent-going-anywhere-so-what-does-that-mean-for-todays-teens.

140 ***Wall Street Journal* report:** Wells, Georgia, Jeff Horwitz, and Deepa Seetharaman. "Facebook Knows Instagram Is Toxic for Teen Girls, Company Documents Show." *Wall Street Journal*, September 14, 2021. https://www.wsj.com/articles/facebook-knows-instagram-is-toxic-for-teen-girls-company-documents-show-11631620739.

141 **Another study, reported:** Hughes, Virginia. "Does Social Media Make Teens Unhappy? It May Depend on Their Age." *New York Times*, March 28, 2022. https://www.nytimes.com/2022/03/28/science/social-media-teens-mental-health.html.

142 **went mega-viral in 2015:** Hollis, Rachel. "Bikini Picture." *Rachel Hollis* (blog), March 31, 2015. https://msrachelhollis.com/2015/03/31/bikini-picture/.

143 **"Birthday Q&A" post:** Covington, Caitlin. "Birthday Q&A!" *Southern Curls & Pearls* (blog), July 2018. https://www.southerncurlsandpearls.com/birthday-q/#/.

143 **Caitlin decided to open up:** Kelly, Kristen. "Caitlin Covington Says Social Media Fame Caused Crippling Anxiety: 'I Wanted to Stay in Bed All Day.'" *People*, December 2, 2019. https://people.com/health/caitlin-covington-says-social-media-fame-caused-crippling-anxiety/#:~:text=Her%20posts%20began%20garnering%20more,the%2028%20year%20old%20says.

148 **gave birth to Kennedy:** Covington, Caitlin. "Kennedy's Birth Story." *Southern Curls & Pearls* (blog), March 2021. https://www.southerncurlsandpearls.com/kennedys-birth-story/.

149 **after Kennedy's birth:** Covington, Caitlin (cmcoving). "40 weeks pregnant vs. 1 week postpartum. . . ." Instagram post, January 24, 2021. https://www.instagram.com/p/CKc0HRDAJ6y/?hl=en.

150 **keep it real:** Covington, Caitlin (cmcoving). Instagram Story. https://www.instagram.com/cmcoving.

151 **candid blog post:** Covington, Caitlin. "3 Months With Kennedy." *Southern Curls & Pearls* (blog), April 2021. https://www.southerncurlsandpearls.com/3-months-with-kennedy/#/.

Chapter 7

157 **Instagram in January 2021:** Valerio, Mirna (themirnavator). "Nothing and no space is apolitical. . . ." Instagram video, January 8, 2021. https://www.instagram.com/reel/CJzIPRBn2kk/?hl=en.

159 **announcement on Facebook:** Valerio, Mirna. "I hope you and yours are doing well. . . ." Facebook, July 9, 2020. https://www.facebook.com/TheMirnavator/posts/pfbid033BVoqn6HgYRDsPiWsjdsaeCsxo6MxxS9HUMsyTcdcCbWZzWfDud6CzAVwcuuWw8gl.

160 **"New Face of Fitness":** Hiland, Sophie. "Vermont's New Face of Fitness: The Mirnavator." VTSports.com, August 7, 2020. https://vtsports.com/the-mirnavator/.

160 ***Seven Days* magazine:** Edgar, Chelsea. "Montpelier Endurance Athlete and Advocate Mirna Valerio Is Taking Up Space." *Seven Days*, August 12, 2020. https://www.sevendaysvt.com/vermont/montpelier-endurance-athlete-and-advocate-mirna-valerio-is-taking-up-space/Content?oid=30984058.

161 **the Influencer League:** MSL. "MSL Study Reveals Racial Pay Gap in Influencer Marketing," December 6, 2021. Cision. https://www.prnewswire.com/news-releases/msl-study-reveals-racial-pay-gap-in-influencer-marketing-301437451.html.

161 **"These are stark numbers":** Brown, Stacey, and NNPA. "Study Reveals Racial Pay Gap for Social Media Influencers." The Atlanta Voice, December 29, 2021. https://theatlantavoice.com/study-reveals-racial-pay-gap-for-social-media-influencers.

162 **digital marketing agencies:** Warren, Jillian. "Nano or Macro: How an Influencer's Follower Count Impacts Engagement Rate." *Later* (blog), April 9, 2021. https://later.com/blog/influencer-engagement-rate/.

166 **series of Instagram Stories:** Lage, Ayana (ayanagabriellelage). "What's the point of posting publicly in support of #BlackLivesMatter? . . ." Instagram video, June 1, 2020. https://www.instagram.com/tv/CA5kBochqK-/?hl=en.

171 **named Cara Dumaplin:** McNeal, Stephanie. "This Popular Parenting Influencer Is a Trump Donor. What Should Followers Do about It?" *BuzzFeed News*, January 22, 2021. https://www.buzzfeednews.com/article/stephaniemcneal/taking-cara-babies-trump-donations-controversy.

173 **colleague at *BuzzFeed News*:** Mack, David. "The Women in the 'Christian Girl Autumn' Meme Want You to Know Something." *BuzzFeed News*, August 13, 2019. https://www.buzzfeednews.com/article/davidmack/christian-girl-autumn-meme-caitlin-convington-emily-gemma.

173 **"If all of Twitter":** Covington, Caitlin (@cmcoving)."If All of Twitter Is Gonna Make Fun of My Fall Photos, at Least Pick Some Good Ones! Super Proud of These. for the Record, I Do like Pumpkin Spice Lattes. Cheers! Pic.twitter.com/Qzflqtwqae." Twitter, August 12, 2019. https://twitter.com/cmcoving/status/1160980974483775490?lang=en.

175 **by her views:** "Influencer Discussion, Tuesday Sept 29." Reddit. Accessed October 20, 2022. https://www.reddit.com/r/blogsnark/comments/j1wi8d/influencer_discussion_tuesday_sept_29/.

177 **"for me to do so":** Anonymous. "Influencer Discussion, Thursday Feb 24." Reddit, February 24, 2022. https://www.reddit.com/r/blogsnark/comments/t07coo/influencer_discussion_thursday_feb_24/.

Chapter 8

185 **mommy blog obsession:** Mendelsohn, Jennifer. "Honey, Don't Bother Mommy. I'm Too Busy Building My Brand." *New York Times*, March 12, 2010. https://www.nytimes.com/2010/03/14/fashion/14moms.html.

185 **"Of all those kids":** BirdBrain Parenting Skills. "Birdalamode/Shannon Bird." *Get off My Internets* (blog), May 1, 2021. https://gomiblog.com/forums/beauty-fashion-bloggers/bird-ala-mode-shannon/page-694/.

186 **"24 hours ago Shannon":** quietbright. "Shannon Bird." Reddit, 2020. https://www.reddit.com/r/blogsnark/comments/hplz7z/shannon_bird/.

188 **hours and hours on YouTube:** Perez, Sarah. "Kids Now Spend Nearly as Much Time Watching TikTok as YouTube in US, UK and Spain." *TechCrunch*, June 4, 2020. https://techcrunch.com/2020/06/04/kids-now-spend-nearly-as-much-time-watching-tiktok-as-youtube-in-u-s-u-k-and-spain/.

189 **A Change.org petition:** "Remove Myka Stauffer's MONETIZED YouTube Videos Exploiting a Special Needs Child." Change.org. Accessed October 20, 2022. https://www.change.org/p/youtube-remove-myka-stauffer-s-monetized-youtube-videos-exploiting-a-special-needs-child.

193 **actors are treated:** Lambert, Harper. "Why Child Social Media Stars Need a Coogan Law to Protect Them from Parents." *Hollywood Reporter*, August 20, 2019. https://www.hollywoodreporter.com/business/digital/why-child-social-media-stars-need-a-coogan-law-protect-parents-1230968/.

193 **the labor union:** "Coogan Law." Sagaftra.org. Accessed October 20, 2022. https://www.sagaftra.org/membership-benefits/young-performers/coogan-law.

194 **a child can work:** "Child Entertainment Laws as of January 1, 2022." United States Department of Labor. Accessed October 20, 2022. https://www.dol.gov/agencies/whd/state/child-labor/entertainment#:~:text=Yes-,Sec.,employment%20in%20the%20entertainment%20industry.&text=Title%2026%20Sec.,they%20must%20have%20work%20permits.

194 **writes Marina Masterson:** Masterson, Marina. "When Play Becomes Work: Child Labor Laws in the Era of 'Kidfluencers.'" *University of Pennsylvania Law Review* 169, no. 2 (2021): 577–607.

195 **"largely unworkable in the fast-paced social media context":** Masterson. "When Play Becomes Work."

196 **"Under this 'kidfluencer' bill"**: Lambert. "Why Child Social Media Stars Need a Coogan Law to Protect Them from Parents."

197 **In 2020, French lawmakers:** McNeal, Stephanie. "The French Are Stepping Up to Protect Kids Making Money Online." *BuzzFeed News*, October 9, 2020. https://www.buzzfeednews.com/article/stephaniemcneal/french-pass-law-protecting-kid-influencers.

197 **"right to be forgotten"**: AFP, Le Monde avec. "Le Parlement Adopte à L'unanimité Une Loi Pour Encadrer La Pratique Des Enfants 'Influenceurs.'" *Le Monde*, October 6, 2020. https://www.lemonde.fr/societe/article/2020/10/06/le-parlement-adopte-a-l-unanimite-une-loi-pour-encadrer-la-pratique-des-enfants-influenceurs_6055019_3224.html.

Chapter 9

205 **Super Bowl of blogging:** McNeal, Stephanie. "Here's Everything You Wanted to Know about Influencers and the Nordstrom Anniversary Sale." *BuzzFeed News*, August 12, 2020. https://www.buzzfeednews.com/article/stephaniemcneal/nordstrom-rewardstyle-anniversary-sale-questions-answered.

Chapter 10

223 **TikTok is gaining ground:** Bain, Phoebe. "For Influencer Marketing, Instagram Isn't the Only Show in Town Anymore." *Marketing Brew*, October 11, 2022. https://www.marketingbrew.com/stories/2022/10/11/for-influencer-marketing-instagram-isn-t-the-only-show-in-town-anymore.

223 **over the past decade:** Mosseri, Adam. "A Creator-Led Internet, Built on Blockchain." Filmed April 2022 in Vancouver, BC. TED video, 13:37. https://www.ted.com/talks/adam_mosseri_a_creator_led_internet_built_on_blockchain/transcript.

224 **June 2021 blog post:** Mosseri, Adam. "Shedding More Light on How Instagram Works." *Instagram Blog*, June 8, 2021. https://about.instagram.com/blog/announcements/shedding-more-light-on-how-instagram-works.

224 **who feel differently**: McNeal, Stephanie. "These Influencers Are Quitting Instagram, and They Want Others to Join Them." *BuzzFeed News*, February 2, 2022. https://www.buzzfeednews.com/article/stephaniemcneal/influencers-quitting-instagram.

236 **"world around you"**: Valerio, Mirna (themirnavator). "What a weird time to be celebrating Global Running Day. . . ." Instagram post, June 3, 2020. https://www.instagram.com/p/CA_YwZxnE64/.

Chapter 11

240 **"The bubble is starting to burst":** Kapner, Suzanne, and Sharon Terlep. "Online Influencers Tell You What to Buy, Advertisers Wonder Who's Listening." *Wall Street Journal*, October 20, 2019. https://www.wsj.com/articles/online-influencers-tell-you-what-to-buy-advertisers-wonder-whos-listening-11571594003.

240 **according to one report:** MarketsandMarkets. "Influencer Marketing Platform Market Worth $24.1 Billion by 2025—Exclusive Report by MarketsandMarkets™." Cision, December 17, 2020. https://www.prnewswire.com/news-releases/influencer-marketing-platform-market-worth-24-1-billion-by-2025-exclusive-report-by-marketsandmarkets-301195104.html.

245 **Girl Scouts Gold Award:** Stiffler, Lisa. "Kids Are Non-Consenting Stars of Some Family 'Vlogs'—and a High Schooler Wants to Change That." *GeekWire*, February 15, 2022. https://www.geekwire.com/2022/kids-are-non-consenting-stars-of-some-family-vlogs-and-a-high-schooler-wants-to-change-that/.